Springer Theses

Recognizing Outstanding Ph.D. Research

For further volumes:
http://www.springer.com/series/8790

Aims and Scope

The series "Springer Theses" brings together a selection of the very best Ph.D. theses from around the world and across the physical sciences. Nominated and endorsed by two recognized specialists, each published volume has been selected for its scientific excellence and the high impact of its contents for the pertinent field of research. For greater accessibility to non-specialists, the published versions include an extended introduction, as well as a foreword by the student's supervisor explaining the special relevance of the work for the field. As a whole, the series will provide a valuable resource both for newcomers to the research fields described, and for other scientists seeking detailed background information on special questions. Finally, it provides an accredited documentation of the valuable contributions made by today's younger generation of scientists.

Theses are accepted into the series by invited nomination only and must fulfill all of the following criteria

- They must be written in good English.
- The topic should fall within the confines of Chemistry, Physics, Earth Sciences, Engineering and related interdisciplinary fields such as Materials, Nanoscience, Chemical Engineering, Complex Systems and Biophysics.
- The work reported in the thesis must represent a significant scientific advance.
- If the thesis includes previously published material, permission to reproduce this must be gained from the respective copyright holder.
- They must have been examined and passed during the 12 months prior to nomination.
- Each thesis should include a foreword by the supervisor outlining the significance of its content.
- The theses should have a clearly defined structure including an introduction accessible to scientists not expert in that particular field.

Manuela Mura

Self-Assembly of Flat Organic Molecules on Metal Surfaces

A Theoretical Characterisation

Doctoral Thesis accepted by King's College London, UK

Springer

Author
Dr. Manuela Mura
University of Central Lancashire Preston
Lancashire
UK

Supervisor
Prof. Lev Kantorovich
King's College London
London
UK

ISSN 2190-5053 ISSN 2190-5061 (electronic)
ISBN 978-3-642-30324-1 ISBN 978-3-642-30325-8 (eBook)
DOI 10.1007/978-3-642-30325-8
Springer Heidelberg New York Dordrecht London

Library of Congress Control Number: 2012940851

Printed on acid-free paper

Springer is part of Springer Science+Business Media (www.springer.com)

Part of this thesis have been published in the following journal articles

- F. Silly, A.Q. Shaw, M.R. Castell, G.A.D Briggs, M. Mura, N. Martsinovich and L.N. Kantorovich. J. Phys. Chem C 2008, **112**, 11476. "Melamine Structures on the Au(111) Surface".
- M. Mura, N. Martsinovich and L.N. Kantorovich. Nanotech. 2008, **19**, 46704. "Theoretical study of melamine super-structures and their interaction with the Au(111) surface".
- M. Mura, A. Gulans, T. Thonhauser and L.N. Kantorovich. Phys. Chem. Chem. Phys. 2010, **12**, 4759. "Role of van derWaals interaction in forming molecular-metal junctions: flat organic molecules on the Au(111) surface".
- M. Mura, F. Silly, G.A.D Briggs, M.R. Castell and L.N. Kantorovich. J. Phys. Chem C 2009, **113**, 21840. "H-Bonding supramolecular assemblies of PTCDI molecules on the Au(111) surface".
- M. Mura, X. Sun, T.H. Jonkman, F. Silly, G.A.D Briggs, M.R. Castell and L.N. Kantorovich. Phys. Rev. B 2010, **81**, 195412. "Experimental and theoretical analysis of H-Bonding supramolecular assemblies of PTCDA molecules on the Au(111) surface".
- W.Xu, J. Wang, M.F. Jacobsen, M. Mura, M. Yu, R.E.A. Kelly, Q. Meng, E. Laegsgaard, j. Kjems, T.R. Linderoth, L.N. Kantorovich, K. Gothelf and F. Besenbacher. Angewandte Chemie 2010, **49**, 9373 "Supramolecular Porous Network Formed by Molecular Recognition between Chemically Modified Nucleobases G and C".
- M. Mura, F. Silly, Burlakov, V. G.A.D Briggs, M.R. Castell and L.N. Kantorovich. Science. "Formation mechanism for a hybrid supermolecular network involving cooperative interactions". submitted.

Supervisor's Foreword

The thesis of Manuela Mura is a theoretical study of the self-assembly of a collection of organic molecules on a gold surface. This is a very exciting area in modern theoretical and experimental research, which lies on a boundary of surface science and nanotechnology: on the one hand, adsorption of molecules on a crystal surface is a domain of surface science; on the other hand, molecules assemble into intriguingly ordered networks which can be utilised for various applications, and these become the remit of nanotechnology. Indeed, nanotechnology deals with nanostructures, which are designed to serve a particular useful purpose (or function), such as, e.g. sensor particular molecules in a gas or forming an artificial material with predefined properties. One way of constructing nanostructures is of course to build them step-by-step from its elements, i.e. atoms or molecules, using surface probe methods such as scanning tunnelling and atomic force microscopies (STM and AFM). However, even if we knew what to build and how, it is an extremely difficult (although becoming increasingly possible!) and tricky procedure. The main problem, however, is that this method cannot be utilised in manufacturing when many devices are needed to be built. Self-assembly may be the answer since molecules arrange into beautiful, highly ordered structures themselves, without human intervention, and large structures covering significant surface areas can be grown in this way. The problem of course is that one has to learn how the desired structure is to be built. Although nature does the work, one has to show her the way. We need to learn how to affect the structures of the molecular assemblies, how to direct their growth in a particular direction without, e.g. ending up with a collection of needed and not so much polymorphs. The coin has two sides: thermodynamics and kinetics. The former is about free energies of the assemblies, while the latter—about the way the assemblies actually grow under particular experimental conditions. Although much work has been done over the last 10 years on the thermodynamic aspect of self-assembly, very little is known about their kinetics. In order to solve all these problems, a combined approach of theory and experiment is absolutely essential with both feeding and inspiring each other. Experiment and theory look at the same problem from different perspectives, and this helps in tackling it. And this thesis is an example of such a collaborative approach since all systems studied here theoretically were done jointly with experimental groups in Nottingham and Oxford (both in the United Kingdom) and Aarhus (in Denmark). The thesis contains a systematic theoretical study of a number of small organic molecules on the gold surface, which form various planar two-dimensional assemblies by means of hydrogen bonding. All essential facets of the process are covered, from understanding how and why the molecules bind to each other and to the surface, to their mobility and kinetics of growth. For small molecules such as Melamine, which was used throughout as a testing ground, many assemblies in the gas phase (i.e. without considering the surface) are possible, and a systematic approach of constructing all possible structures is proposed. Then, interaction with the surface is considered. For the first time, dispersion (or van der Waals) interaction between molecules and the gold surface was accounted for in a self-consistent way using recently developed first-principles density functional, and these calculations allowed better understanding of the binding and mobility of planar organic molecules on the gold surface; on top of that, an existing controversy related the observed evaporation temperatures for molecules of different sizes was also resolved. The developed methodology is then applied to several other molecules on the same surface, and the reader will see a great variety and complexity of the structures they form, and how theory helped in understanding them in great detail. Finally, first steps towards elucidating kinetics aspects of Melamine self-assembly are made, especially related to trapping native gold atoms in the hexagonal pores of the network they form. The fascinating point here is that gold atoms cannot penetrate the growing assembly due to insurmountable energy barriers; hence, the only possibility was to assume that some specific mechanism is at play that "delivers" gold atoms into the pores of the network during its growth, and such a 'gating' mechanism is proposed. I sincerely hope that this book will be of great help and inspiration for young scientists working in the field of self-assembly of molecules on crystal surfaces.

Lev Kantorovich

Acknowledgments

I would like to express my thanks to:

- My supervisor Prof. Lev Kantorovich, for investing plenty of time and effort making my project a success and his contagious enthusiasm for scientific excellence.
- Dr. Natalia Martsinovich, Dr. Ross Kelly for their help and advice.
- Dr. Fabien Silly, Andris Gulans and Timo Thonauser.
- The EPSRC, for funding the project
- Those who are involved in the basic technology group.
- My parents, brother and the Marco Family.
- Dr. Paolo Ruggerone.
- Prof. Andrei Zvelindovsky for his support.
- All of my friends and colleagues who have made these years in London so wonderful.
- My best friends from Cagliari
- and 'dulcis in fundo' my husband Marco.

Preface

Over the last 10 years the world has been involved in a technological revolution. Every day newer and smaller devices, essential to life, appear on the market: mobile phones, computers, laptops, netbooks and so on. Nowadays, it is possible to have a portable device with 1 Terabyte capacity compared to the 2 Gigabytes being the maximum 4 years ago. This is just an example of how rapid the development in technology can be. The main aim is to decrease the dimensions while increasing the efficiency and capability of the product. In nanotechnology, one is particularly interested in manipulating and designing objects at a nanoscale to be implemented in everyday products. The main aim is to decrease the dimensions while increasing the efficiency and capability of the product. In nanotechnology, one is particularly interested in manipulating and designing objects at a nanoscale to be implemented in everyday products. The organic/ inorganic interfaces are excellent candidates for building electronic devices at a nanoscale, as well as optical sensors.

Developments in nanotechnology are related to the investigation of scanning probe microscopy (SPM) , allowing analysis to be done at the nanoscale (10^{-9} m). With STM they made possible the visualization of surfaces and interfaces by employing the principles of quantum mechanical tunneling. The atomic force microscope (AFM) by using the forces acting between the surface and the tip, an image of the surface or an interface can be obtained. These new techniques make possible an understanding of the physical structure and properties at the nanoscale, changing the focus from the top-down to the bottom-up strategy of building nanodevices.

Experimental methods based on SPM can give invaluable information concerning the templates and the molecular aggregations. However, in many cases atomic or even molecular resolution is not possible, and different theoretical methods will therefore play an important role in resolving the image and modelling the supramolecular templates.

The aim of this thesis is to present a theoretical study of self-assembly of flat organic molecules (such as melamine, PTCDA, PTCDI, NTCDA, NTCDI and DNA derivative molecules) on the Au(111) metal surfac. In order to describe in detail the assembly of the molecules, a systematic approach to building molecular superstructures based on the notion of binding sites has been proposed. First, one must identify all possible sites for hydrogen bonding between molecules. Then how to form molecular pairs and larger structures using all possible combinations of these binding sites has been considered. In this way, all possible dimers, chains and 2D monolayers of melamine, dimers and chains of PTCDA and PTCDI molecules have been considered. The energies of these structures are calculated using the density-functional theory SIESTA code. The strength of hydrogen bonding in various molecular arrangements is analysed. The theoretically predicted monolayer structures are in very good agreement with the results of STM measurements.

Contents

Chapter 1
Introduction

The work in this thesis is focused on molecules that are able to self-assemble on different surfaces by forming two-dimensional templates stabilised via double or triple hydrogen bonding. In particular, assemblies of molecules such as melamine, perylene tetra-carboxylic di-imide (PTCDI), perylene tetra-carboxylic di-anhydride (PTCDA), naphthalene tetracarboxylic-dianhydride (NTCDA) and naphthalene tetracarboxylic diimide (NTCDI) are studied in detail. The aim is to give a complete characterisation of the supramolecular networks, taking into account the balance between the molecule-molecule and molecule-substrate interactions. All our assembly calculations are done within the gas phase approximation, i.e. without taking into account the surface, which is a good approximation assuming that the molecules are quite mobile on the surface. Using a systematic method based on considering all possible hydrogen bond connections between the molecules we investigate planar superstructures that organic molecules can form in one and two dimensions. The structures studied are based on two or more molecules per unit cell and all structures considered, assemble in flat periodic patterns. Most of the calculations are performed using the density functional theory method. We show that the calculated lattice parameters of the structures considered compare well with those measured experimentally. To specifically check the applicability of the gas-phase approximation, we systematically investigated the adsorption of the molecules on the Au(111) metal surface with the particular attention being paid to the characterisation of the potential energy surface of our molecules on this surface. We performed these calculations using both a conventional functional (PBE) which does not include the dispersion interaction, and the newly developed vdW-DF method which does. We find that the adsorption energies of these flat molecules on the metal surface calculated with the vdW-DF method are effected significantly by the dispersion interaction and depend linearly on the size of the molecules. While the PBE method predicts very weak adsorption energies which do not depend on the sizes of the molecules, the vdW-DF method gives strong binding entirely due to the dispersion interaction. We found that both PBE and vdW-DF methods predict a very small corrugation of the total energy of the molecules on gold. These results support our main assumption of

M. Mura, *Self-Assembly of Flat Organic Molecules on Metal Surfaces*,
Springer Theses, DOI: 10.1007/978-3-642-30325-8_1,

the molecule-surface interaction changing little laterally and resulting in a mobility of the molecules at room temperature on the surface, i.e. the gas-phase modelling is a good approximation for the Au(111) surface.

Developments in nanotechnology are related to the investigation of scanning probe microscopy (SPM), allowing analysis to be done at the nanoscale (10^{-9}m). Scanning tunneling microscopy (STM) was first developed by Binning and Rohrer [1, 2]. With STM they made possible the visualization of surfaces and interfaces by employing the principles of quantum mechanical tunneling, and the discovery resulted in a Nobel prize for Physics in 1986 [2]. The atomic force microscope (AFM) was developed in 1986 by Binning, Quate and Geber [3]: by using the forces acting between the surface and the tip, an image of the surface or an interface can be obtained. These new techniques make possible an understanding of the physical structure and properties at the nanoscale, changing the focus from the top-down to the bottom-up [4, 5] strategy of building nanodevices.

Molecular self-assembly is the ability of some molecules to rearrange themselves in ordered templates on crystal surfaces, causing directional, selective, and noncovalent interactions between neighboring molecules [6–8]. These interactions are due to the presence of functional groups that drive the assembly of the molecules through of hydrogen-bonds or metal coordination. However, recent studies also highlight the self assembly for molecules which are able to form complex networks through covalent interactions [9, 10].

This thesis will in particular focus on the stability of self-assembled monolayers strengthened via hydrogen bonding. Despite the weakness of the hydrogen bond, the latter plays a central role in the self-organisation of biological systems [11]. It is responsible for the stabilisation of amino acids pairs in DNA, for the formation of the tertiary (helix or beta-shift), and the quaternary structure (folding) of the proteins [11].

An interesting application is related to the self-assembly of molecules on a surface, where the molecules rearrange themselves in two-dimensional (2D) ordered templates [12]. The interaction between the molecular network and the substrate can be either strong (chemisorption) or weak, driven mainly by *van der Waals* (vdW) forces (physisorption). The balance between the inter-molecular (lateral) and molecule-substrate interactions drives the formation of the supramolecular order. One of the aims of nanotechnology is to explore and control surfaces by forming various assemblies of molecules and atoms on them. In exploring surface phenomena at the nanoscale modern chemistry helps in the creation of new synthetic molecules that can assemble on surfaces. The key point is the introduction, into the molecule, of particular functional groups that can form hydrogen bonds. Playing with the properties of the functional groups and the shape of molecules it is possible to obtain new different templates with well defined properties [4, 5, 12, 13] such as the 2D porous nano-templates of melamine-PTCDA [14], which can trap molecules with well defined functional properties.

Experimental methods based on SPM can give invaluable information concerning the templates and the molecular aggregations. However, in many cases atomic or even molecular resolution is not possible, and different theoretical methods will therefore play an important role in resolving the image and modelling the supramolecular templates. The aim of these methods is to give a complete characterisation of the supramolecular networks, taking into account the balance between the molecule-molecule and molecule-substrate interactions. By using a systematic approach one can characterise the self-assembled molecules in the gas phase (i.e. without taking into account the surface) and then investigate the role played by the substrate in molecular aggregation process.

The present work is focused on molecules able to self-assemble on different surfaces and form hexagonal and square templates by stabilisation through double or triple hydrogen bonds [6, 7, 14, 15]. In particular, assemblies of molecules such as melamine, perylene tetra-carboxylic di-imide (PTCDI) and perylene tetra-carboxylic di-anhydride (PTCDA) will be studied in detail.

A review of recent experimental and theoretical observations on the self-assembly of small molecules is presented in detail in the following section, with particular attention paid to STM experiments performed under ultra-high vacuum (UHV) conditions.

1.1 The Experimental Observations of 2D Supramolecular Monolayers

This section will give an overview of recent literature on self-assembly of small molecules on different surfaces.

As underlined in the previous section, the self assembly of some nano-structures is governed by directional, selective and non-covalent bonds between the molecules. The formation of nano-templates is due to the incorporation of functional groups into the molecules, stabilising the interactions between neighboring molecules through hydrogen bonding. It is well known that the formation of hydrogen bonds is a consequence of the attractive force between one electronegative atom (oxygen, nitrogen or fluorine) and a hydrogen atom covalently bonded to another electronegative atom. Despite its weakness (the associated energy is one orders of magnitude lower than that of a covalent bond), the role of hydrogen-bonding is nevertheless important in biological systems: the secondary structure of DNA, the tertiary (double helix) and the quaternary structure of proteins are due to the formation of hydrogen bonds between different amino-acids and the back-bones of the proteins [11]. Using the SPM techniques one has been able to study and understand the structures of various assemblies and nanotemplates [1–3]. The deposition of organic molecules on a metallic or non-metallic substrate modifies the density of states introducing which may change the electronic behaviour of the "molecule/substrate" system.

Over the last 10–15 years the SPM techniques (see Sect. 1.3) have been used extensively for visualising surface phenomena, such as the molecular arrangement on the surface. Early experiments have been presented, in particular, with organic molecules by depositing nucleic bases and DNA crystals on different surfaces. This deposition highlights the role played by the shape of molecules and the functional group in the assembly of the template. Clearly, the ability to control the molecular ordering on surfaces is a key requirement in producing molecular devices. However, this goal can only be achieved if one can understand the molecular structure of the sample from SPM images.

Recently many experiments have been conducted using SPM to investigate the behaviour of organic molecules deposited on surfaces. The 2D self-assembly of organic molecules leads to a new type of nanostructured template formed by arrays of nano-holes or pores [6, 7, 14, 16, 17]. These structures are interesting as they can be used in the trapping of diffusing molecular species with specific properties [18]. Depending on the nature of the peripheral sites of the molecules involved in the process different networks can be constructed. However, the 2D molecular ordering is a compromise between the adsorbate-adsorbate and adsorbate-substrate interactions. In addition, the deposition of some molecules on surfaces can induce properties, such as chirality, that are not normally observed in the three dimensional (3D) crystal structure. Chirality is a molecular property indicated through mirror-symmetry of the structure and the inability for two opposite chiral structures to be super imposed [5, 9, 13, 19]. STM allows the investigation of the chirality of monolayers adsorbed on substrates. It was possible to observe either a confinement of molecules with the same chirality and the formation of large islands [20] (as is the case with the cyanuric acid and melamine networks) or the formation of complicated structures (guanine adsorption on the Au(111) surface [21]).

Another interesting aspect of these templates is the role of the surface. In some cases the molecules are chemisorbed on the substrate. Molecules such as di-D-phenylalanine are chemisorbed on the Cu(110) surface, whereas 3,4,9,10-perylene-tetracarboxylic-diimide (PTCDI) are chemisorbed on the Ag(111) surface [22]. In these cases the surface is highly active and the geometry of the template is influenced by the presence of the substrate.

In other cases the molecules are physisorbed, i.e. the molecules are mainly bound to the surface through weak dispersion forces. The potential energy surface is quite flat and the molecules are free to move on the surface eventually self-assembling. The periodicity of the assembled network is determined by the molecule–molecule interaction [6–8, 23, 24].

The molecules considered in this thesis are: 1,3,5-triazine-2,4,6-triamine (Melamine), perylene-3,4,9,10-tetracarboxylic-3,4,9,10-dianhydride (PTCDA), 3,4,9,10-perylene-tetracarboxylic-diimide (PTCDI), naphthalene tetracarboxylic anhydride (NTCDA) and naphthalene tetracarboxylic diimide (NTCDI), all shown in Fig. 1.1. These molecules have a particular shape and functionality allowing them to assemble in networks.

1.1.1 Melamine

Melamine molecules are of special interest in engineering of 2D porous nanostructures due to their specific triangular shape and hydrogen-bonding functionality enabling them to serve as vertexes, which can be connected by other long molecules

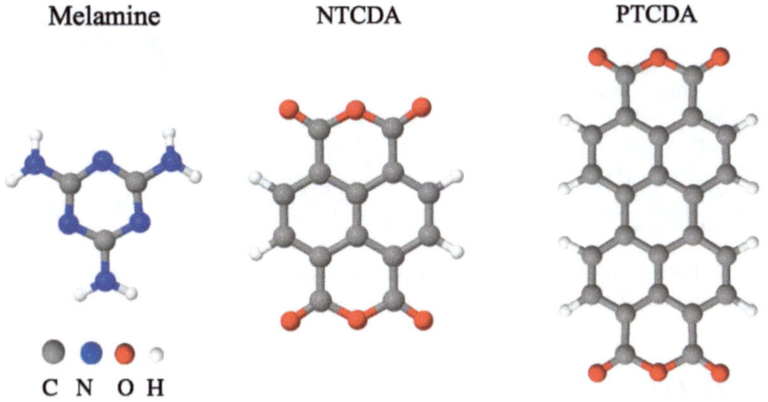

Fig. 1.1 Melamine, NTCDA and PTCDA molecules

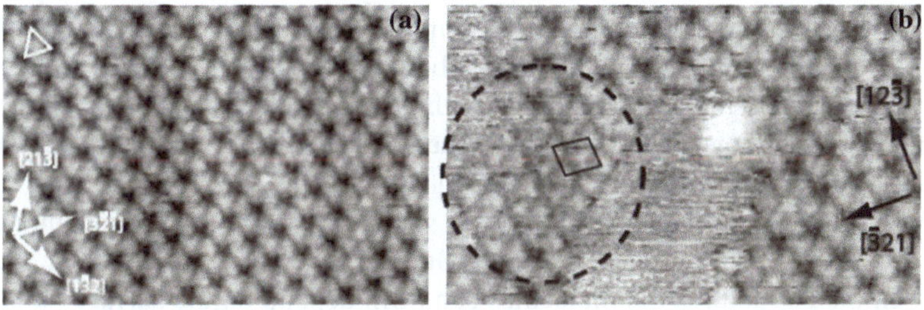

Fig. 1.2 STM image of melamine (**a**) hexagonal and (**b**) square domain on a Au(111)-(22 × $\sqrt{3}$) surface (13 × 9 nm^2; $V_s = -1.1$ V, $I_t = 0.5$ nA [25] [V_s and I_t are the STM parameter for the bias applied and the tunneling current])

into large 2D nanostructures. In particular, complicated supramolecular structures capable of trapping foreign fullerene molecules have been reported [7, 14, 20]. Melamine has been deposited on different surfaces such as Au(111) [18, 25, 26] and Ag/Si(111) [8, 14]. STM studies of melamine on the Au(111) surface show two possible arrangements: a hexagonal [20, 25–27] (Fig. 1.2a) and a closed-packed structure [25] (Fig. 1.2b). This nonchiral molecule is able to form a chiral network [13, 14] on metal surfaces. Moreover, it has the ability to mix with other molecular blocks, such as PTCDA [17, 28], NTCDI [8], PTCDI [6, 14, 29–31] and cyanuric acid [26, 20], creating complex supramolecular architectures with either nonchiral or chiral symmetry.

It is perhaps not immediately obvious from the images that there is no commensurability between the melamine template and the Au(111) surface. The network presents a short range periodicity, where the dimer is repeated to form a hexagonal or square pattern, and a long range periodicity [20] due to the non commensurability between the network and the surface.

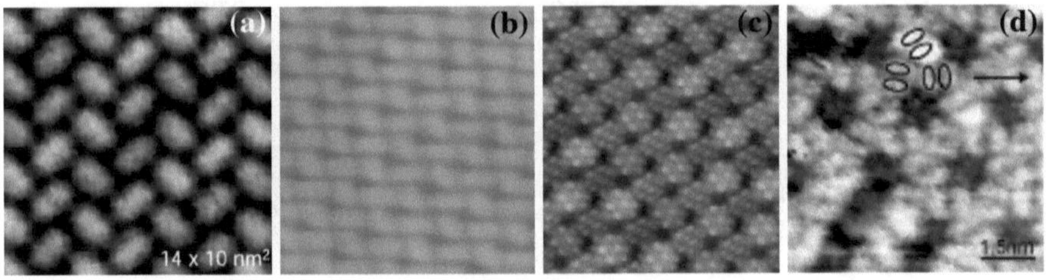

Fig. 1.3 STM images of the four PTCDA phases: **a** herringbone phase [16], **b** square phase [16], **c** brick wall [34] and **d** hexagonal phase [16]

1.1.2 PTCDA

Perylene-3,4,9,10-tetracarboxylic-3,4,9,10-dianhydride (PTCDA) molecules, see Fig. 1.1, are of special interest in engineering of 2D porous nanostructures due to their special shape and a special hydrogen bonding functionality. These long molecules are employed, alone or combined with other molecules such as melamine, as building blocks of complicated supramolecular structures, or templates, capable of trapping foreign molecules (e.g. fullerenes [14–17, 20]).

PTCDA molecules have been largely studied because they have interesting chemical and physical properties, such as fluorescence and semi-conductance. The self-assembled structures the molecules can form depend on the surface nature and preparation, as well as the deposition rate and temperature. PTCDA molecules were deposited on metallic Cu(110) [32], Cu(111) [33], Ag(110) [34], Ag(111) [34, 35], Au(100) [36], Au(788) [37] and Au(111) [24, 28, 36–40], non-metallic KBr(001) [41, 42], NaCl [43], on a bilayer of graphene deposited on the SiC(0001) surface [44], GaAs(001) [45], Si/Ag(111) [16] and mica [46]. Despite the considerable effort, only four molecular structures have been observed so far: (i) herringbone [16, 24, 28, 34–37, 41, 42], see Fig. 1.3a, (ii) square [16, 24, 32, 36, 37], see Fig. 1.3b, (iii) brick wall [34, 44], see Fig. 1.3c and (iv) a more complex hexagonal phase [16] which can be considered as a combination of a herringbone and square phases, see Fig. 1.3(a, b). Finally a new arrangement has been observed, the "domino" phase, in Fig. 1.4b. In the domino phase the molecules are arranged in squares, each square is made of four molecules attached to each other as domino pieces.

The structures described above were observed on different surfaces, i.e., the herringbone phase was also observed by depositing PTCDA on the sulphur passivated GaAs(001) [45], see Fig. 1.5.

The PTCDA on the Cu(110) [32] surface has a low coverage (0.6 monolayer (ML) immediately after deposition of molecules at room temperature) and the mobility of molecules was low. After thermal annealing to 450 K the PTCDA rearrange as in Fig. 1.6. By depositing PTCDA on the Cu(111) [33] two possible arrangements were observed, see Fig. 1.7. These two phases have been called 1a and 2a.

The deposition of PTCDA on Ag(110) [34] and Ag(111) [34, 35] resulted in two different phases. PTCDA deposited on Ag(110) surface assemble in a square phase,

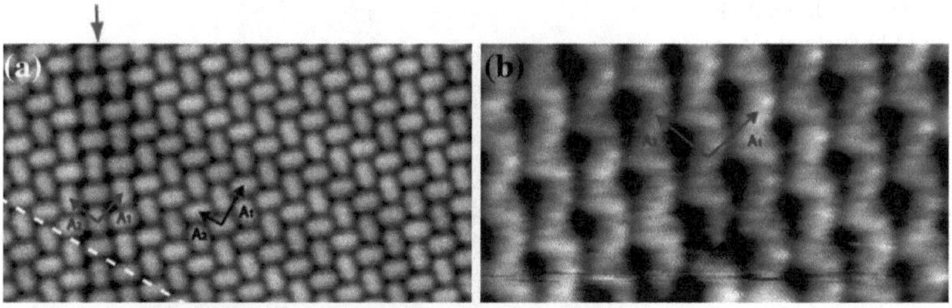

Fig. 1.4 STM images of the observed new PTCDA networks on the Au(111) surface. **a** The narrow stripe of the square phase providing a buffer for two shifted herringbone phases on the *left* and *right* of it. The *big black arrow* at the top of the image indicates the square phase. The two lattice vectors of the two square and herringbone phases are explicitly indicated by *blue* and *black arrows*, respectively. The scanned area is 210 × 150 Å^2. The tunneling current $I_t = 0.3$ nA and the applied voltage $V_s = 1.5$ V. **b** The domino phase; the two lattice vectors are indicated by the *blue arrows*. The scanned area is 105 × 70Å^2. The tunneling current $I_t = 0.7$ nA and the applied voltage $V_s = -0.3$ V [47]

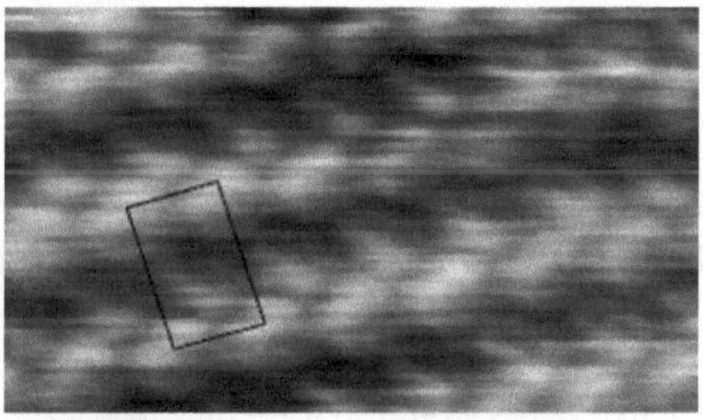

Fig. 1.5 The herringbone arrangement of PTCDA molecule on the S-GaAs (001) surface ($I_t = 0.08$ nA, $V_s = -3.3$ V, deposition of PTCDA at 180 °C) [45]

Fig. 1.6 PTCDA deposited on the Cu(110) surface annealed to 450 K (I_t = 300 pA, $V_s = -0.45$ V) [32]

see Fig. 1.8a, while the deposition on the Ag(111) surface is observed the usual herringbone structure, see Fig. 1.8b. There are few groups that studied the deposition of PTCDA on the Au(111) surface [24, 36, 38–40]. The STM image of PTCDA on Au(111) shows one herringbone arrangement and a square phase (Fig. 1.9).

Some experiments have been conducted by depositing PTCDA on the KBr(001) surface [41, 42] and on NaCl [43] in both the cases the herringbone phase has been observed. Other STM experiments have been done on SiC [44], where the herringbone structure (80 K, $V_s = 2.1$ V, $I_t = 4.8$ pA) and the brick-wall structure (4.7 K, $V_s = 1.2$ V, $I_t = 16$ pA) have been observed. Interesting the deposition of PTCDA

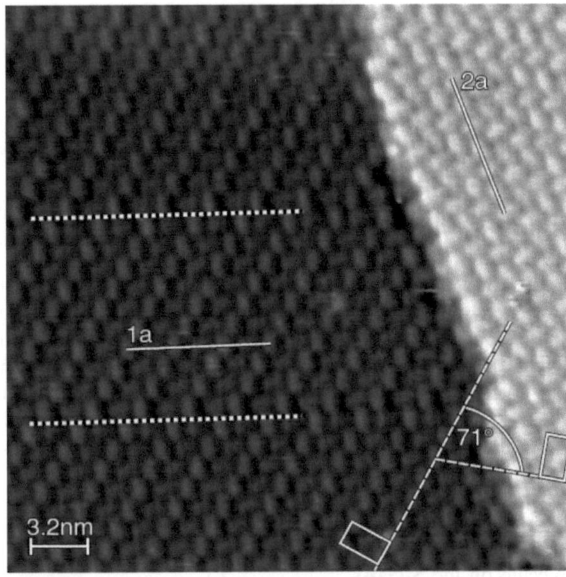

Fig. 1.7 PTCDA deposited on the Cu(111) surface ($V_s = -0.9\,\text{V}$, $I_t = 3.3\,\text{pA}$) [33]

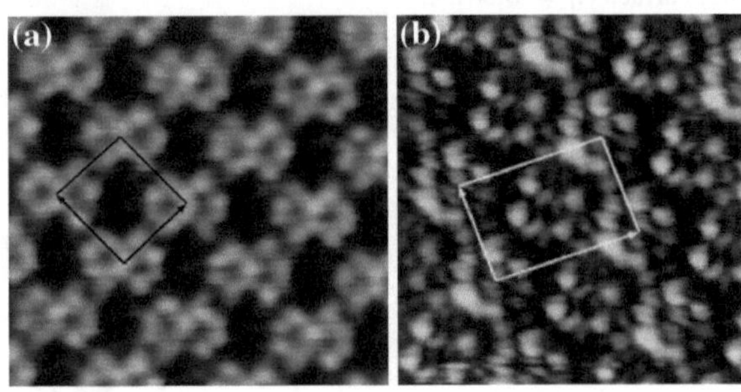

Fig. 1.8 **a** STM image of the PTCDA monolayer on the Ag(110) surface ($V_s = 0.90\,\text{V}$ and $I_t = 1.0\,\text{nA}$) [34] and **b** PTCDA monolayer on the Ag(111) surface ($V_s = 0.20\,\text{V}$ and $I_t = 1.3\,\text{nA}$) [35]

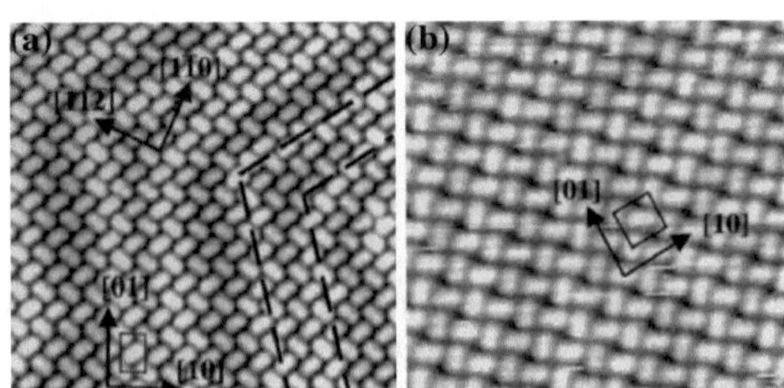

Fig. 1.9 STM images of the herringbone phase **a** ($V_s = -1.44\,\text{V}$, $I_t = 0.11\,\text{nA}$ at room temperature) and the square phase ($Vs = -1.29\,\text{V}$, $I = 0.1\,\text{nA}$) of PTCDA on the Au(111)-($23 \times \sqrt{3}$) surface [24]

on the Si-Ag(111) surface [16] evidence three different phases: the herringbone ($V_s = 2.0\,\text{V}$, $I_t = 0.1\,\text{nA}$), the square ($V_s = -2.5\,\text{V}$, $I_t = 0.1\,\text{nA}$) and another more complex phase called the hexagonal phase ($V_s = -3.0\,\text{V}$, $I_t = 0.1\,\text{nA}$), see Fig. 1.3d.

1.1.3 PTCDI

Perylene-tetracarboxylic-diimide (PTCDI) molecules, see Fig. 1.1, are shaped like PTCDA, but have different chemical properties due to different hydrogen bonding functionality at their shorter sides.

From the experimental point of view, PTCDI have been studied less than PTCDA molecules. As far as we are aware, these molecules were deposited only on the graphite and MoS_2 [48], Ag/Pt(111) [49], Pt(100) [50], NaCl(001) [51], Au(111) [29, 52, 53] and Si/Ag(111) [16] surfaces, and basically three structures has been reported: canted, brick-wall and domino [53]. In the canted structure the molecules sandwiched between the parallel rows are rotated and are connected by their edges [16, 52, 54]. In Fig. 1.10a we see rows of canted molecules running parallel to each other [52, 16]. Molecules in every other row are tilted in the same direction at an angle which is approximately 12 ° to the row direction. The shapes of the molecules in the rows between these ones are not well resolved. However, there are indications that these are tilted in the opposite direction by 12 °. We believe that this is a canted structure already observed on the Au(111) and Ag/Si(111) surfaces previously [16, 29, 52]. In the "domino" structure shown in Fig. 1.10b the molecules are arranged in squares, each square made out of four molecules attached to each other as domino pieces. Two different contrasts are observed in the centre of the squares delimited by the PTCDI molecules. Larges squares appear brighter than small ones. The bare gold surface is observed in the large square, whereas the PTCDI interactions in the small squares lead to a local decrease of the density of states. Similar phenomenon was observed in melamine network on Au(111), where melamine hexagon centre appears lower than the gold surface in the STM images [25]. In the structure seen in Fig. 1.10c molecules are arranged in identical rows shifted with respect to each other; molecules within rows are aligned along the rows direction. This arrangement is equivalent to the brick-wall structure formed by PTCDA molecules on a number of substrates [34, 44].

A lot of experiment conducted on the Ag(111) surface indicate that the type of bonding is likely to be physisorption [22].

1.1.4 NTCDA and NTCDI

Naphthalene tetracarboxylic-dianhydride (NTCDA) and Naphthalene tetracarboxylic diimide (NTCDI) are two molecules that have a similarity with PTCDA and PTCDI molecules, respectively, described above, see Fig. 1.1. They have the same termination on both ends of the NTCDA/NTCDI molecules, which may form hydrogen bonds with melamine. The main difference between NTCDA, NTCDI and PTCDA, PTCDI is the distance between the two ends of molecules: NTCDA, NTCDI molecules are much shorter.

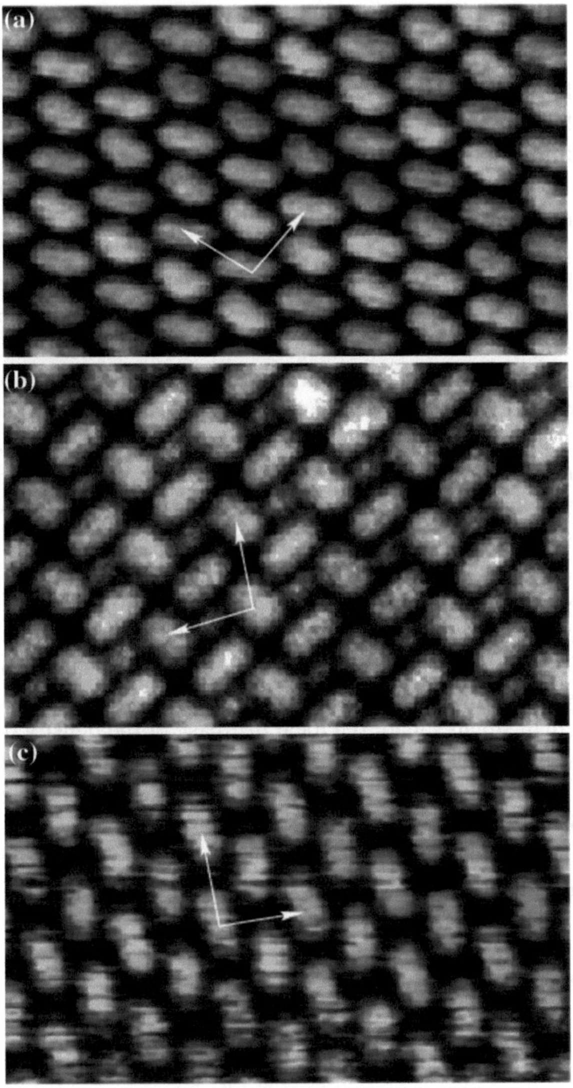

Fig. 1.10 Three PTCDI monolayers observed with STM on the Au(111) surface: canted (**a**), domino (**b**) and brick-wall (**c**) structures. The lattice vectors are indicated by *white vectors* in each case. Tunneling parameters: (**a, b**) $I_t = 0.1\,\text{nA}$, $V_s = -1.8\,\text{V}$; (**c**) $I_t = 0.5\,\text{nA}$, $V_s = -1.8\,\text{V}$

The deposition of NTCDA on the Si/Ag(111) [12] results in the formation of diffuse islands which nucleated at the step edges and in the defect terminations, see Fig. 1.11a. At the same time deposition of NTCDA on the Au(111) surface [55] results in a herringbone structure.

The deposition of NTCDI on the Si/Ag(111) surface [12] results in the formation of single rows and close-packed islands of NTCDI molecules [8, 12, 56], see Fig. 1.11b. The NTCDI deposited on the Au(111) surface [56] assemble into a

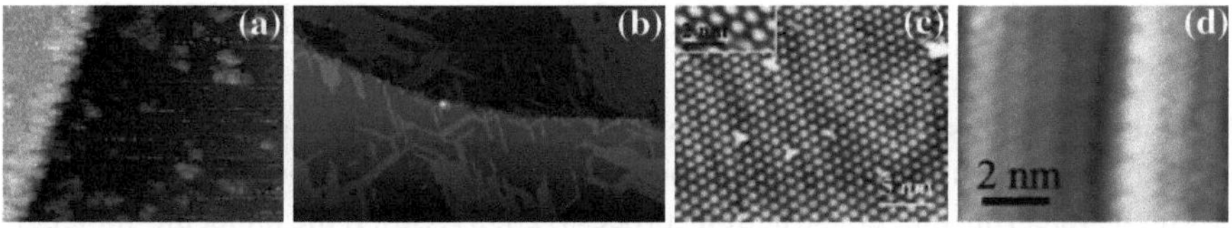

Fig. 1.11 **a** The $50 \times 65\,\text{nm}^2$ STM image ($I_t = 0.1$ nA, $V_s = -3.0$ V) of NTCDA adsorbed on the Ag/Si(111)$\sqrt{3} \times \sqrt{3}\,R30°$ surface [12] where there are the formation of diffuse islands; **b** the $95 \times 165\,\text{nm}^2$ STM image ($I_t = 0.05$ nA, $V_s = -1.2$ V) of NTCDI adsorbed on the Ag/Si(111)$\sqrt{3} \times \sqrt{3}R30°$ surface [12]; **c** 13.4×11.4 Å^2 STM image ($I_t = 0.162$ nA, $V_s = -1.6$ V) of NTCDI adsorbed on the Au(111) surface [56] and **d** 24.3×15.4 AA^2 STM image ($I_t = 0.162$ nA, $V_s = -1.6$ V) of NTCDI adsorbed on the Au(455) surface [56]

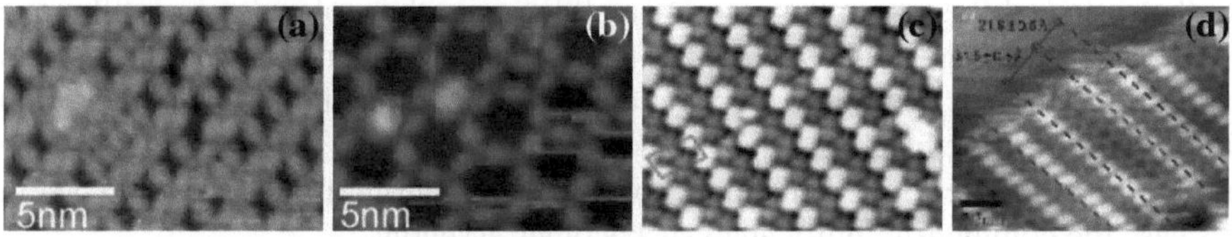

Fig. 1.12 STM images of the structures formed after depositing PTCDI and melamine molecules on the Au(111) surface: **a** the parallelogram network ($V_s = -1.80\,V$, $I_t = 0.03$ nA, $17.0\,\text{nm} \times 14.0\,\text{nm}$); **b** the hexagonal network ($V_s = -1.80\,V$, $I_t = 0.03$ nA, $15.1\,\text{nm} \times 12.4\,\text{nm}$) [18]; **c** mixed PTCDA and melamine domain ($V_s = -1.50$ V, $I_t = 0.01$ nA, $14\,\text{nm} \times 8\,\text{nm}$) [28]; **d** mixed PTCDI and melamine domain ($V_s = 1.50$ V, $I_t = 0.03$ nA, $32.9\,\text{nm} \times 33.9\,\text{nm}$) [18]

herringbone (Fig. 1.11c) structure while on the Au(455) surface [56] a close-packed structure is observed (Fig. 1.11d).

1.1.5 Mixed Phases PTCDA/PTCDI-Melamine

Experimentally assembly into interesting networks was observed when these molecules are deposited on the surface. Structures such as the parallelogram and hexagonal domain, see Fig. 1.12a and b, but also other mixed phases with different composition of melamine and PTCDA/PTCDI molecules were seen, (Fig. 1.12c and d), the hexagonal phase may exist alongside the parallelogram network [17]. These phases often form from the deposition of PTCDA and melamine [17, 28] and PTCDI melamine [7, 14, 18] on the Au(111) and the Ag-Si(111) surfaces. The melamine/PTCDI ratio for this mixed phases is 2:1 for the structure shown in Fig. 1.12c and 4:1 for the structure shown in Fig. 1.12d, while for the hexagonal and parallelogram phases the ratio is 2:3. Therefore, in order to work out the periodicity of this structures it is necessary to consider unit cells with more than 2 molecules.

1.2 Theoretical Issues Related to 2D Supramolecular Monolayers

Particular structures formed during the deposition depend crucially on the interplay between the molecule–molecule and molecule-surface interactions which determine the kinetics steering the "construction" process [28, 57, 58]. If the molecule-molecule interaction dominates, then molecules may move around the surface to form molecular assemblies (providing, of course, that the surface diffusion is not hampered by the too large deposition rate [13, 59, 60]; in this case commensurability issues do not arise and the *gas-phase* modeling of the assemblies [61–65], i.e. without considering the surface, is justified. If, however, the molecule-surface interaction is much stronger than their mutual interaction, then all depends on the actual energy barriers for the surface diffusion: in the case of small barriers, the molecules can still move around to form the assemblies steered by inter-molecular interactions; if the barriers are significant, then one has to start the analysis by finding all possible adsorption sites. In the intermediate case the full calculation involving the assembly and the surface should be performed and the whole potential energy surface analyzed, i.e. in this case the commensurability analysis becomes essential.

Density Functional Theory (DFT) has become an invaluable tool in studying molecular assemblies (see, e.g. [23, 25, 62–75]). It is essential, in the present case, to have an accurate characterisation of the kind of interaction (hydrogen bond) stabilised between the single molecules that are involved in the assembly. The strength of the hydrogen bond is dependent on the electronegativity of the hydrogen donor and acceptor atoms. There are three types of hydrogen bonds: strong, moderate and weak. This classification is strongly dependent on the electronegativity of donor and acceptor atom and the geometrical formation of the whole complex donor–acceptor. We are mainly interested in strong hydrogen bond. Previous *ab-initio* calculations done for DNA bases investigate the accuracy of the DFT method using different density functionals in the description of the hydrogen bond formation [76–79] and compared the results with more accurate Møller-Plesset perturbation theory (MP2) studies [80–83]. One can use these calculations as the benchmarks to identify the strength of the hydrogen-bonds and justify the choice of the PBE functional to characterise them. In these calculations the results obtained for different hydrogen bonds with PBE and BLYP GGA functionals calculated using the SIESTA method, PBE functional using the VASP method and MP2 method were compared. They show a good performance of the PBE functional compared with the more accurate MP2 approach [80–82].

The main concern is to understand the role played by the vdW interaction in formations of hydrogen bonded assemblies of simple flat molecules on the surface. The Au(111) surface was chosen since it represents a popular platform for looking for flat ordered molecular arrangements: it can be easily prepared with relatively large terraces available, is well studied, and allows performing Scanning Tunneling Microscopy (STM) experiments. In many cases, e.g. for DNA bases [64, 65, 75, 84, 85] and melamine molecules on the Au(111) surface [23, 25], PTCDA molecules on the Ag(111) surface [67, 69, 70], benzene [73] and PTCDA

[86] molecules on the Au(111), Cu(111) and Ag(111) surfaces the calculated adsorption energies are too small (of the order of 0.1–0.2 eV) which is inconsistent with experimental observations of the desorption temperatures for these molecules (around 100–300 °C). This failure of standard generalized gradient approximation (GGA) functionals, such as e.g. Perdew–Burke–Ernzerhof (PBE) [87, 88], is attributed to the incorrect description of the electron correlation in situations when the electronic densities of the molecule and surface practically do not overlap and the dispersion interaction, or van der Waals (vdW) forces, become dominant [73, 89]. The local density approximation (LDA) typically predicts stronger binding than GGA, also for covalent systems, however, this effect is artificial [90, 91].

The second order MP2, coupled-cluster and other methods of quantum chemistry are amongst the available methods used to study systems for which the dispersion forces is important. Unfortunately, in many cases, their application are limited to small systems. The questions about the main binding mechanism of the molecules to stick to the surface and the values of the binding energies were addressed in [23, 25, 64, 75] for some DNA bases on the gold surface using a semiempirical approach based on the correction to the PBE energies by the vdW interaction obtained in Ref. [92] by fitting to the second order MP2 calculations of gold clusters with fragments of nucleic base molecules. However, since the DFT is successfully used in the physics community to describe systems there is the necessity to improve description of dispersion forces with this method. There are two main methodologies available that are able to treat the vdW interaction using first-principles calculations. One approach consists in adding to the standard DFT an additional semiempirical term which takes into account the dispersion forces (DFT-D) [93, 94] while the second one is a truly first-principles approach which does not require any fitting parameter (vdW-DF) [95]. Both methods have been successfully applied to a large variety of systems. They were used to calculate the interaction energies as a function of the distance for the benzene dimer in the sandwich and two T-shape configurations [93, 96]. In both the cases the results obtained are in good agreement with quantum chemistry calculations [97].

The DFT-D approach is implemented in a large number of *ab-initio* codes and it has been accurately tested for carbon and organic based systems [93, 94, 98–100] and recently has been tested for molecule-metal interface systems as well [101]. On the other hand the vdW-DF is a rather new technique that has only been applied to a limited number of systems; however, this method has an advantage as being a truly first-principles basedmethod that has not got any parameters to adjust [91, 96, 102–104].

1.3 Experimental Techniques STM and AFM

In order to control the formation of nanostructures with desired properties, methods are needed capable of characterisation of their structure. This is provided by two techniques: Scanning Tunnelling Microscopy (STM) and Atomic Force Microscopy

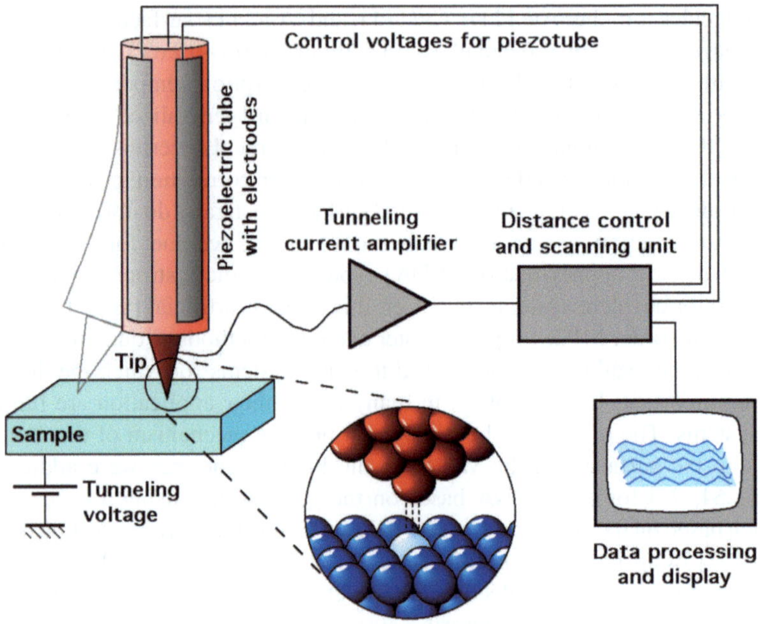

Fig. 1.13 Schematic STM apparatus [105]

(AFM). These allows the imaging the surface structures and are classified as Scanning Probe Microscopy (SPM) techniques. STM was invented in 1981 by Binning and Heinrich Rohrer [1, 2], while the AFM was invented by Binning, Quate and Gerber [3] in 1986.

These techniques enable one to investigate the surface phenomena, and to understand the principle factors of supra-molecular ordering in two dimensions and control the outcome of the self-assembly process itself. STM and AFM exploit different quantum physical properties of the molecule-surface-tip junction. Which physical properties do they exploit?

1.3.1 STM

STM uses the tunneling effect between a metallic tip and a conductive surface. An experimental STM set-up (Fig. 1.13) can be schematically represented as a metallic tip brought very close to a conductive surface. When the voltage is applied the tunneling effect occurs between the probe and the sample, as shown in Fig. 1.13. The tip is mounted on a piezoelectric tube allowing for tip tiny movements by applying a voltage. This guarantees a very high precision in the change of its lateral and vertical positions with respect to the surface. When a bias is applied between the tip and the sample a net intensity of current is measured. The scan can be done at constant tunneling current (constant current mode) which makes it possible to visualise the

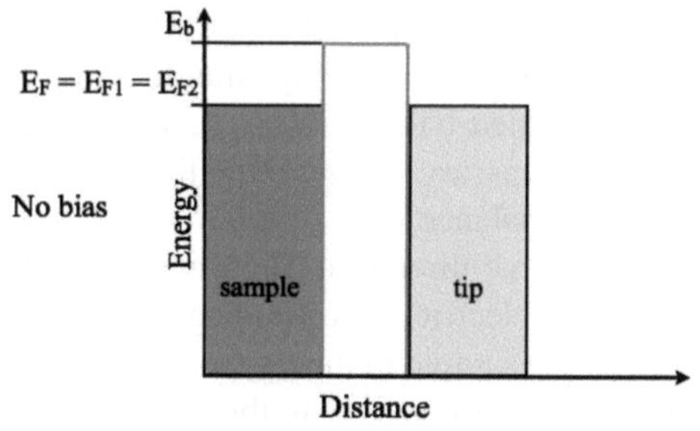

Fig. 1.14 Schematic representation of the tip and the sample when the bias is not applied, but two are electronically connected (i.e. Fermi level alignment)

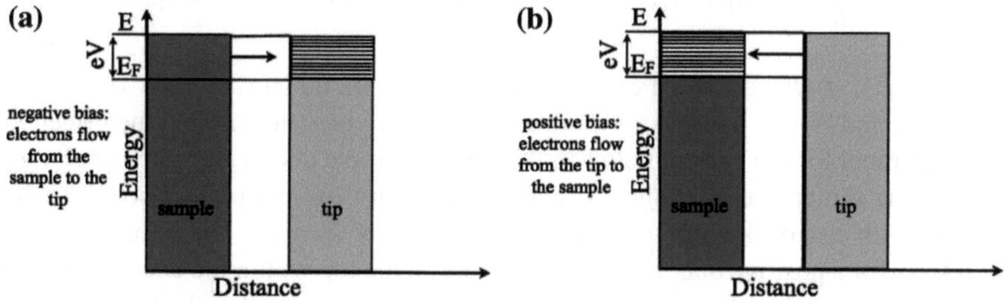

Fig. 1.15 Schematic representation of the electronic levels of the tip and the sample when a negative (**a**) or positive (**b**) bias, with respect to the surface, is applied

topography of the surface. The topographic image approximately is the spatial map of the local density of states affect by the exponential dependence of the tunneling current (I_t) on the tip-surface distance [1, 2]. In another approach the tip keeps its position at a constant height (constant-height mode) with the tunneling current measured across the surface. In the constant current mode the STM image could be considered as a map of the tip height at different positions across the surfaces. The same argument can be repeated in the constant height mode, because the height is fixed but the actual tunneling current is measured across all surface.

A schematic overview of the physical principles lying at the heart of the STM is presented in Figs. 1.14 and 1.15.

The tunneling from tip to the sample (and vice versa) is well described by quantum mechanics as a transfer of one electron through a single potential barrier. The latter is the vacuum between the tip and the sample. In Fig. 1.14 the electronic levels are schematically represented as filled colour boxes. E_{F1} and E_{F2} are the Fermi levels of the tip and of the sample, respectively. Above the Fermi energy all the levels are unoccupied. In the case of no applied bias (Fig. 1.14) and assuming that Fermi energies of the tip and the sample are equal ($E_{F2} = E_{F1}$), no current flows from one media to the other. Quantum mechanics asserts that there is a small but non-zero probability that an electron will cross or tunnel through the "forbidden" vacuum gap. The resulting current is proportional to the tunneling probability. The latter is related to the corresponding transmission coefficient $T(E)$ where E is the electron energy. Unfortunately, the flow takes place in two opposite directions and there will be no net current (I) due to the detailed balance ($I_{\text{sample}\rightarrow\text{tip}} = I_{\text{tip}\rightarrow\text{sample}}$). In order to have a tunneling effect it is necessary to overcome the barrier (E_b) by applying the bias voltage. As a result, the balance is destroyed ($I_{\text{sample}\rightarrow\text{tip}} \neq I_{\text{tip}\rightarrow\text{sample}}$) and there is a net current (I) in some direction. The direction of the tunneling current depends on the polarity of the electric field. If a negative bias respect the surface is applied electrons flow from the sample to the tip (see Fig. 1.15, left panel), but when a positive bias is applied, the electrons flow in the opposite direction (see Fig. 1.15, right panel). The net current through the junction "surface-tip" is then given by

$$I_{t\rightarrow s} = \frac{2\pi e}{\hbar} \int |M_{ts}|^2 N_t(E - eV) N_s(E) f_t(E - eV)[1 - f_s(E)] dE \qquad (1.1)$$

where $f_s(E)$, $f_t(E)$ are the Fermi-Dirac functions of the surface and tip, $E' = E + |e|\,V$, and $N_t(E)$ and $N_s(E)$ denote their density of states. M_{ts} is the effective matrix element for the current between the tip and sample due to the states of energy E, and V is the applied voltage. At zero temperature the Fermi functions can be replaced with a step-like function which equals to 1 for $E < E_F$ and 0 for $E > E_F$. Therefore, for a positive bias $V > 0$, only states on the top within the energy interval $E_F < E < E_F + |e|\,V$ would contribute to the current and the electrons will be transmitted into empty states of the surface, while for negative bias $V < 0$, the surface states, $E_F - |e|\,V < E < E_F$ would contribute to the current with surface electrons transmitted into unoccupied states of the tip.

The current depends strongly on the transmission probability. The current is also very sensitive to the tip position and decreases exponentially with an increasing tip-sample distance. This makes it important to the control of the lateral and vertical positions of the tip with respect to the sample. A simple calculation of the STM image was suggested by Bardeen [106], which is based on time-dependent perturbation theory. In this approach the key assumption is that the interaction between the tip and the sample is very weak. The current will depend on the amount of overlap between the wave-functions of both electrodes, which gives the tunnelling matrix element, M. Using Fermi's golden rule it is possible to derive the probability of one electron tunneling from one electrode into another. The matrix element has the same physical meanings as the transmission probability. Tersoff and Hamann [107] simplified the Bardeen formula even further assuming that the tip wave-function is spherical, in which case the formula for the current would not longer depend on the tip structure at all, only the density of states of surface and tip position.

The STM technology can be employed to study different kinds of materials such as: metals or semiconductors, and molecules on conductive surface including biological molecules (e.g. DNA).

The advantage of STM is a high atomic resolution, but unfortunately the chemical structure of the tip is unknown; since the results of the scan do depend on the tip structure, their interpretation may be not straightforward; in addition this technique can be used only to study conductive materials. Also, in many cases, including

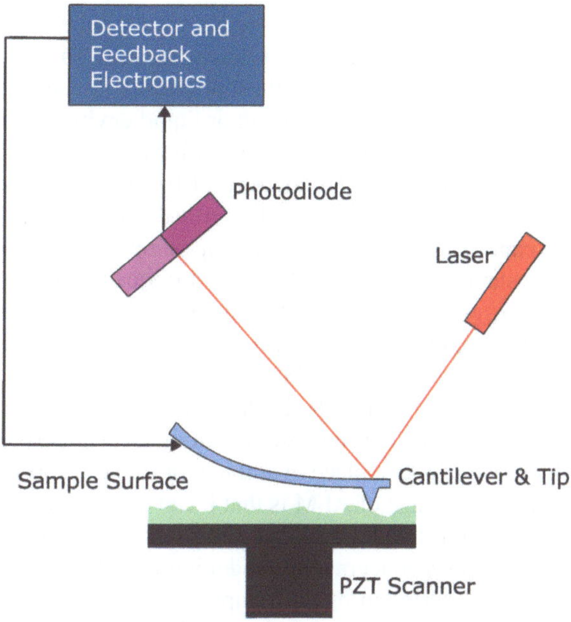

Fig. 1.16 A vertical deflection of the cantilever is detected by reflecting a laser beam off the back of the cantilever. The photodiode signal is used to drive a feedback system which controls the movement of the piezo xyz excitation signal and oscillation frequency. The change in the cantilever deflection is due to the tip-sample interaction force [108]

(especially organic) molecules on surfaces atomic resolution has not been achieved, like in the brick-wall phase of PTCDA shown in Fig. 1.3c, or the brick-wall phase of PTCDI on the Au(111) surface, Fig. 1.10c.

Most of the STM experiments are conducted under ultra high vacuum (UHV), however, STM in a solution or ambient is also possible.

1.3.2 AFM Technique

The Atomic Force microscopy (AFM) technique is based on exploiting the forces between tip and sample and monitoring the surface under study.

As in the case of STM, Sect. 1.3, in the AFM apparatus there is a tip that scan the sample under study. The AFM apparatus, schematically shown in Fig. 1.16, consists of a microscale cantilever which ends with a sharp tip (probe) that is used to scan the surface. The cantilever is typically made of silicon or silicon nitride and the tip has a radius of curvature of the order of tens of the nanometers.

The forces measured in AFM in UHV are due to long range (electrostatic and *van der Waals*) and short range forces (chemical bonds).

Typically, the deflection of the tip is measured using a laser spot reflected from the top of the cantilever into an array of photodiodes. There are different possible regimes of operation of the AFM set-up: static (contact), dynamic (non-contact) and intermittent or tapping AFM. The contact mode operates by scanning the tip across the sample surface while the change in cantilever deflection is monitored. A feedback loop maintains a constant deflection between the cantilever and the sample for each (x, y) data point while the tip scan the sample. The vertical position of the tip at each (x, y) point is stored to form the topographic image of the sample. Operation can take place either in vacuum, ambient or liquid environments.

In the non-contact mode the cantilever oscillates slightly above the surface at a resonance frequency (amplitude is of 10 nm). The tip is not in contact with the sample surface, but oscillates above the surface during scanning. The resonant frequency of the cantilever is a function of the tip-sample force. Several feedback loops maintain a constant oscillation amplitude and frequency by vertically moving the scanner at each (x, y) data point and adjusting the excitation signal during the scanning. In the AFM tapping mode the cantilever oscillates at near resonance constant frequency (amplitudes are typically from 20 to 100 nm). The tip lightly taps on the sample surface during scanning. The feedback loops maintain a constant oscillation amplitude by maintaining a constant oscillation signal acquired by the split photodiode detector. Operation can take place in ambient or liquid environments.

The advantage of AFM is that it can analyse semiconductors, metals and insulators, and it has high versatility and simplicity of use and high resolution.

This thesis concerns the modelling of the STM images observed. Therefore, an accurate description of the computational methods used to perform the theoretical work and a theoretical description of the STM approach we used to perform simulations will be given in the next chapter. In the following Chapters we shall present the results reached, performing the calculations. In Chap. 3 we shall describe the molecule-molecule interaction in the gas phase and the comparison of derived gas-phase structures with the corresponding STM images. In Chap. 4 we shall investigate in the molecule-substrate interaction and the role played by the surface in the assemblies of the molecules. Chapter 5 is dedicated to examining the presence/absence of "bright spots" in the melamine hexagonal network due to trapped gold atoms.

Most of the work presented in this thesis has been published. The gas phase work presented in Chap. 3 on the melamine structures have been published in [23, 25, 109], the PTCDA work in [47, 109], on the PTCDI in [53, 109] and on the NTCDA/NTCDI molecules in [109]. The results presented in Chap. 4 have been published for the melamine in [23] and for PTCDA and NTCDA in [109]. The results presented in Chap. 6 for the DNA derivatives have been submitted in [110].

References

1. Binnig G, Rohrer H, Gerber C, Werbel E (1983) Phys Rev Lett 50:120
2. Binnig G, Rohrer H (1987) Rev Mod Phys 59:615
3. Binnig G et al (1985) Phys Rev Lett 55:991
4. De Feyter S, De Schryver FC (2003) Chem Soc Rev 32:139
5. De Feyter S, Miura A, Yao A, Chen Z, Wurthner F, Jonkheijm P, Schenning APH, Meijer EW, De Schryver FC (2005) Nano Lett 5:77
6. Perdigão LMA, Champness NR, Beton PH (2006) Chem Commun 5:538
7. Perdigão LMA, Perkins EW, Ma J, Staniec PA, Rogers BL, Champness NR, Beton PH (2006) J Phys Chem B 110:12539

8. Perdigão LMA, Fontes GN, Rogers BL, Oxtoby NS, Goretzki G, Champness NR, Beton PH (2007) Phys Rev B 76:245402
9. Lingenfelder M, Tomba G, Constantini G, Colombi Ciacchi L, De vita A, Kern K (2007) Angew Chem Int Ed 46:4492
10. Tomba G, Lingenfelder M, Constantini G, Kern K, Klappernberger F, Barth JV, De Vita J (2007) Phys Chem A 111:12740
11. Berg JM, Tymoczko JL, Stryer L (2002) Biochemistry. Michelle Julet, New York
12. Keeling DL, Oxtoby NS, Wilson C, Humphry MJ, Champness NR, Beton PH (2003) Nano Lett 3:129
13. Barth JV, Costantini G, Kern K (2005) Nature 437:671
14. Theobald JA, Oxtoby NS, Phillips MA, Champness NR, Beton PH (2003) Nature 424:1029
15. Ma J, Rogers BL, Humphry MJ, Ring DJ, Goretzki G, Champness NR, Beton PH (2006) J Phys Chem B 110:12539
16. Swarbrick JC, Ma J, Theobald JA, Oxtoby NS, O'Shea JN, Champness NR, Beton PH (2005) J Phys Chem B 109:12167
17. Swarbrick JC, Rogers BL, Champness NR, Beton PH (2006) J Phys Chem B 110:6110
18. Saywell A, Magnano G, Satterley CJ, Pedigão LMA, Champness PH, Beton NR, O'Shea JN J Phys Chem C
19. Barth JV (2007) Annu Rev Phys Chem 58:375
20. Staniec PA, Perdigão LMA, Rogers BL, Champness NR, Beton PH (2007) J Phys Chem C 111:886
21. Xu REA, Kelly W, Gersen H, Lægsgaard E, Stensgaard I, Kantorovich LN, Besenbacher F (2009) Prochiral guanine adsorption on au(111): An entropy-stabilized intermixed guanine-quartet chiral structure. Small 5:1952
22. Sassi M, Oison V, Debierre J (2008) Surf Sci 602:2862
23. Mura M, Martsinovich N, Kantorovich LN (2008) Nanotechnology 19:465704
24. Chinzhov I, Kahn A, Scoles G (2000) J Crystal Growth 208:449
25. Silly F, Shaw AQ, Castell MR, Briggs GAD, Mura M, Martsinovich N, Kantorovich LN (2008) J Phys Chem C 112:11476
26. Xu W, Dong M, Gersen H, Rauls E, Vázquez-Campos S, Crego-Calama M, Reinhoudt DN, Stensgaard I, Laegsgaard E, Linderoth TR, Besenbacher F (2007) Small 3:854
27. Zhang H, Xie Z, Long L, Zhong H, Zhao W, Mao B, Xu X, Zhao W (2008) J Phys Chem C 112:4209
28. Silly F, Weber UK, Shaw AQ, Burlakov VM, Castell MR, Briggs GAD, Pettifor DG (2008) Phys Rev B 77:201408
29. Silly F, Shaw AQ, Briggs GAD (2008) Chem Comm 16
30. Silly F, Shaw AQ, Briggs GAD, Castell MR (2008) Appl Phys Lett 92:023102
31. Silly F, Shaw AQ, Pettifor DG, Briggs GAD, Castell MR (2007) Appl Phys Lett 91:253109
32. Gabriel M, Stöhr M, Möller R (2002) Appl Phys A 74:303
33. Wanger Th, Bannani A, Bobisch C, Karacuban H, Moller R (2007) Condens Matter 19:056009
34. Glockler K, Seidel C, Sokolowski M, Umbach E, Bohringer M, Berndt R, Schneider W-D (1998) Surf Sci 405:1–20
35. Kilian L, Hauschild A, Temiroc R, Soubatch S, Scholl A, Reinert F, Lee T-L, Tautz FS, Sokolowski M, Ulbach E (2008) Phys Rev Lett 100:136103
36. Mannsfeld S, Toerker M, Schmitz-Hubsch T, Sellam F, Fritz T, Leo K (2001) Org Electron 2:121
37. Kröger J, Jensen H, Berdt R, Rurali R, Lorente N (2007) Chem Phys Lett 438:249
38. Schmitz-Hübsch T, Fritz T, Sellam F, Staub R, Leo K (Mar 1997) Epitaxial growth of 3,4,9,10-perylene-tetracarboxylic-dianhydride on au(111): a stm and rheed study. Phys Rev B 55(12):7972–7976
39. Fenter P, Schreiber F, Zhou L, Eisenberger P, Forrest SR (Aug 1997) In situ studies of morphology, strain, and growth modes of a molecular organic thin film. Phys Rev B 56(6):3046–3053
40. Nicoara N, Romani E, Gomez-Rodriguez JM, Martin-Gago J, Mendez J (2006) Org Electr 7:287

41. Kunstmann T, Schlarb A, Fendrich M, Wangner Th, Moller R, Hoffmann R (2005) Phys Rev B 71:121403
42. Fendrich M, Kunstmann T, Paulkowski D, Möller R (2007) Nanotechnology 18:084004
43. Burke SA, Ji W, Mativetsky JM, Topple JM, Fostner S, Gao H-J, Guo H, Grutter P (2008) Phys Rev Lett 100:186104
44. Lauffer P, Emtsev KV, Graupner R, Seyller T, Ley L (2008) Phys Stat Sol 245:2064
45. N. Nicoara N, Custance O, Granados D, Garcia JM, Gomez-Rodriguez JM, Baro AM, Mendez J (2003) Phys Condens Matter 15:S2619
46. Forker R, Dienel T, Fritz T, Muller K (2006) Phys Rev B 74:165410
47. Mura M, Sun X, Jonkman HT, Silly F, Briggs GAD, Castell MR, Kantorovich L (2010) Experimental and theoretical analysis of h-bonded supramolecular assemblies of ptcda molecules on the au(111) surface. Phys Rev B (in print)
48. Ludwig C, Gompf B, Petersen J, Strohmaier R, Eisenmenger W (1994) Stm investigations of ptcda and ptcdi on graphite and mos2. W Z Phys B 93:365–373
49. Ait-Mansour K, Treier M, Ruffieux P, Bieri M, Jaafar R, Groning P, Fasel R, Groning O (2009) Template-dierected molecular nanostrutures on the ag/pt(111) dislocation network. J Phys Chem C 113:8407–8411
50. Guillermet O, Glachant A, Hoarau JY, Mossoyam JC, Mossoyam M (2004) Perylene tetracarboxylic diimide ultrathin film depositon on pt(1 0 0): a leed, aes and stm study. Surf Sci 548:129–137
51. Topple JM, Burke SA, Fostner S, Grütter P (2009) Thin film evolution: dewetting dynamics of a bimodal molecular system. Phys Rev B 79(20):205414
52. Cañas ME, Xiao W, Wasserfallen D, Müller K, Brune H, JV Barth, Fasel R (2007) Angew Chem Int Ed 46:1814–1818
53. Mura M, Silly F, Briggs GAD, Castell MR, Kantorovich L (2009) H-bonding supramolecular assemblies of ptcdi molecules on the au(111) surface. J Phys Chem C 113:21840–21848
54. Nowakoski R, Seidel C, Fuchs H (2004) Surf Sci 562:53–64
55. Ziroff J, Gold P, Bendounan A, Forster F, Reinert F (2000) Surf Sci 603:354
56. Silanes I, Ruiz-Oses M, Gonzalez-Lakunza N et al (2006) Self-assembly of heterogeneous supramolecular structures with uniaxial anisotropy. J Chem Phys B 110(51):25573–25577
57. Taylor JB, Beton PH (2006) Phys Rev Lett 97:236102
58. Weber UK, Burlakov VM, Pedigão LMA, Fawcett RHJ, Beton PH, Champness NR, Briggs GAD, Pettifor DG (2008) Phys Rev Lett 100:156101
59. Stepanow S, Lingenfelde M, Dmitriev A, Spillmann H, Delvigne E, Lin N, Deng X, Cai C, Barth JV, Kern K (2004) Nature 3:229
60. Kühnle A (2009) Curr Opin Colloid Interface Sci 14:157
61. Perdigão LMA, Staniec PA, Champness NR, Kelly REA, Kantorovich LN, Beton PH (2006) Adenine monolayers on ag-terminated si(111). Phys Rev B 73:195423
62. Otero R, Lukas M, Kelly REA, Xu W, L<E6>gsgaard E, Stensgaard I, Kantorovich LN, Besenbacher F (2008) Elementary structural motifs in a random network of cytosine adsorbed on a gold(111) surface. Science 319:312–315
63. Kelly REA, Kantorovich LN (2006) Planar nucleic acid base super-structures. J Mater Chem 16:1894–1905
64. Kelly REA, Xu W, Lukas M, Otero R, Mura M, Lee YJ, Lægsgaard E, Stensgaard I, Kantorovich LN, Besenbacher F (2008) An investigation into the interactions between self-assembled adenine molecules and the au(111) surface. Small 4:1494
65. Xu W, Kelly REA, Schöck M, Otero R, Laesgaard E, Stensgaard I, Kantorovich LN, Besenbacher F (2007) Probing the hierarchy of thymine-thymine interactions in self-assembled structures by manipulation with scanning tunneling microscopy. Small 3: 2011–2014
66. Heimel G, Romaner L, Brédas J-L, Zojer E (2006) Phys Rev Lett 96:196806
67. Picozzi S, Pecchia A, Gheorghe M, Di Carlo A, Lugli P, Delly B, Elstner E (2003) Phys Rev B 68:195309

68. Hauschild A, Karki K, Cowie BCC, Rohlfing M, Tautz FS, Sokolowski M (2005) Phys Rev Lett 94:036106
69. Hauschild A, Karki K, Cowie BCC, Rohlfing M, Tautz FS, Sokolowski M (2005) Phys Rev Lett 95:209602
70. Rurali R, Lorente L, Ordejon P (2005) Phys Rev Lett 95:208601
71. Seitsonen AP, Lingenfelder M, Spillmann H, Dmitriev A, Stepanow S, Lin N, Kern K, Barth JV (2006) J Am Chem Soc 128:5634
72. Kelly REA, Kantorovich LN (2005) Hexagonal adenine networks constructed from their homo-pairings. Surf Sci 589:139–152
73. Bilic A, Reimers JR, Hush NS, Hoft RC, Ford MJ (2006) J Chem Theory Comput 2:1093
74. Riben M, Payer D, Landa A, Comisso A, Gattinoni C, Lin N, Collin J-P, Sauvage A, De Vita J-P, Kern K (2006) J Am Chem Soc 128:15644
75. Kelly REA, Lukas M, Kantorovich LN, Otero R, Xu W, Mura M, Laesgaard E, Stensgaard I, Besenbacher F (2008) Understanding disorder of the dna base cytosine on the au(111) surface. J Chem Phys 129:184707
76. Kelly REA, Kantorovich LN (2006) J Mater Chem 16:1894
77. Kelly REA, Lee YJ, Kantorovich LN (2005) J Phys Chem B 109:11933
78. Kelly REA, Lee YJ, Kantorovich LN (2006) J Phys Chem B 110:2249
79. Kelly REA, Lee YJ, Kantorovich LN (2005) J Phys Chem B 109:22045
80. Sponer J, Leszczynski J, Hobza PJ, Phys Chem
81. Sponer J, Leszczynski J, Hobza PJ, Mol Str (Theochem)
82. Sponer J, Leszczynski J, Hobza P, Biopolymers
83. Jurecka P, Sponer J, Cemý J, Hobza P (2009) Phys Rev B 79:201105
84. Otero R, Xu W, Lukas M, Kelly REA, Laegsgaard E, Stensgaard I, Kjems J, Kantorovich L, Besenbacher F (2008) Specificity of watson-crick base pairing on a solid surface studied at the atomic scale. Angew Chem Int Ed 47:9673–9676
85. Lukas M, Kelly R, Kantorovich L, Otero R, Laesgaard E, Stensgaard I, Besenbacher F (2009) Adenine monolayers on the au(111) surface: structure identification by stm experiment and ab initio calculations. J Chem Phys 130:024705
86. Romaner L, Nabok D, Puschnig P, Zojer C, Ambrosch-Draxl E (2009) N J Phys 11:053010
87. Perdew JP, Burke K, Ernzerhof M (1998) Phys Rev Lett 80:891
88. Perdew JP, Burke K, Ernzerhof M (1996) Phys Rev Lett 77:3865
89. Priya S, Puschnig P, Nabok D, Ambrosch-Drax C (2007) Phys Rev Lett 99:176401
90. Wu X, Vargas S, Nayak MC, Lotrich V, Scoles G (2001) J Chem Phys 115:8748
91. Langreth DC, Lundqvist BI, Chakarova-Draxl SD, Cooper VR, Dion M, Hyldgaard P, Kelkkanen A, Kleis J, Kong LZ, Li S, Moses PG, Murray E, Puzder A, Rydberg H, Schroder E, Rydberg T, Thonhauser H (2009) Cond Matter 21:084203
92. Piana S, Bilic A (2006) J Phys Chem B 110:23467
93. Grimme S (2004) J Comput Chem 25:1463
94. Grimme S (2006) J Comput Chem 27:1787
95. Langreth DC, Dion M, Rydberg H, Schroder E, Hyldgaard P, Lundqvist BI (2005) J Quantum Chem 101:599
96. Thonhauser T, Puzder A, Langreth DC (2006) J Chem Phys 124:164106
97. Jurecka P, Sponer J, Cemý J, Hobza P (2006) Phys Chem Chem Phys 8:1985
98. Piacenza M, Grimme S (2005) J Am Chem Soc 127:14841
99. Morgado C, Vincent MA, Hillier IH, Shan X (2007) Phys Chem Chem Phys 9:448
100. Foster ME, Shohlberg K (2010) Phys Chem Chem Phys 12:307
101. Nguten M-T, Pignedoli CA, Treier M, Fasel R, Paserone D (2010) Phys Chem Chem Phys 12:992
102. Cooper VR, Thonhauser T, Puzder A, Schroeder E, Lundqvist BI, Langreth DC (2008) J Am Chem Soc 130:1304
103. Thonhauser T, Cooper VR, Li S, Puzder A, Hyldgaard P, Langreth DC (2007) Phys Rev B 76:125112
104. Dion M, Rydberg H, Schroeder E, Langreth DC, Lundqvist BI (2004) Phys Rev Lett 92:246401

105. http://en.wikipedia.org/wiki/File:ScanningTunnelingMicroscope_schematic.png
106. Bardeen J (1961) Phys Rev Lett 6:57
107. Tersoff J, Hamann DR (1985) Stm theory. Phys Rev B 31:805
108. http://upload.wikimedia.org/wikipedia/commons/thumb/7/7c/Atomic_force_microscope_block_diagram.svg
109. Mura M, Gulans A, Thonhauser T, Kantorovich L (2009). Role of van der waals interaction in forming molecule-metal junctions: flat organic molecules on the au(111) surface. Phys Chem Chem Phys (submitted)
110. Xu W, Wang J, Jacobsen MF, Mura M, Yu M, Kelly REA, Meng Q, Laegsgaard E, Kjems J, Linderoth TR, Kantorovich L, Gothelf KV, Besenbacher F. Supramolecular porous network formed by molecular recognition between chemical modified nucleobases g ans c. Angew Chem Int Ed

Chapter 2
Theoretical Methods

Many of the properties of the solid state and chemical systems can be determined solving the Schrödinger equation for a given system. However, states of most of electrons and nuclei have to be accounted for. The solution to many electrons can be obtained by the Hartree Fock (HF) method, using the wavefunction of the electrons, or density functional theory (DFT) based methods, using the density function of the electrons instead of solving the Schrödinger equation. The former method is a base for other approaches used in the quantum chemistry community, whereas the latter method has been largely used in the physics community to study the electronic structure of solids. However, over the last 20 years due to the increased efficiency of computers and the accuracy of the DFT functionals, the number of systems studied using DFT method has increased. As a result of this expansion, systems typically studied using quantum chemistry methods, such as organic and inorganic molecules, are being increasingly often studied with DFT methods because of much better efficiency and high quality which is close to that of the quantum chemistry (QC) methods.

Another way to address many body problem is the classical molecular dynamics that is used to derive physical properties of the system from empirical potentials.

2.1 Solving the Schrödinger Equation

To understand the properties of chemical bonding in molecules and extended systems, we consider the Schrödinger equation, which, for a generic system with N_a nuclei of mass M_N and N_e electrons of mass m_e, leads

$$\hat{H}(\mathbf{r}, \mathbf{R})\Psi_a(\mathbf{r}, \mathbf{R}) = E_a\Psi_a(\mathbf{r}, \boldsymbol{R}) \tag{2.1}$$

where $\Psi_a(\mathbf{r}, \mathbf{R})$ is the system many-body wave function, the electronic coordinates are $\mathbf{r} = \{\mathbf{r}_i\}$ and the nuclear coordinates $\mathbf{R} = \{\mathbf{R}_i\}$.

M. Mura, *Self-Assembly of Flat Organic Molecules on Metal Surfaces*,
Springer Theses, DOI: 10.1007/978-3-642-30325-8_2,

The Hamiltonian operator in Eq. (2.1) is expressed by:

$$\hat{H}(\mathbf{r}, \mathbf{R}) = \hat{T}_n + \hat{T}_e + \hat{U}_{n,n} + \hat{U}_{n,e} + \hat{U}_{e,e} \tag{2.2}$$

$$\hat{T}_e = \frac{-\hbar^2}{2m_e} \sum_{i=1}^{N_e} \nabla_i^2 \tag{2.3}$$

$$\hat{T}_n = \frac{-\hbar^2}{2M_N} \sum_{i=1}^{N_a} \nabla_i^2 \tag{2.4}$$

$$\hat{U}_{n,n} = \frac{1}{2} \sum_{j=1} \sum_{j \neq i}^{N_a} \frac{Z_i^* Z_j}{|\boldsymbol{R}_j - \boldsymbol{R}_i|} \tag{2.5}$$

$$\hat{U}_{n,e} = -\sum_{i=1}^{N_e} \sum_{j=1}^{N_a} \frac{Z_j e}{|\boldsymbol{R}_j - \boldsymbol{r}_i|} \tag{2.6}$$

$$\hat{U}_{e,e} = \frac{1}{2} \sum_{i=1}^{N_e} \sum_{j=1, i \neq j}^{N_e} \frac{e^2}{|\boldsymbol{r}_j - \boldsymbol{r}_i|} \tag{2.7}$$

where Z_i and e are the charges of the nuclei and electron respectively, $\hat{T}$ represents the kinetic energy and $\hat{U}$ the potential energy of either nuclei or electrons. Solving Eq. (2.1) exactly for realistic systems is impossible, because of the high number of degrees of freedom involved. Thus, it is required to apply some approximations.

The most important approximation deals with the separation of nuclear and electronic degrees of freedom and it is known as Born-Oppenheimer approximation. In general, electronic and nuclear dynamics are not independent. However, usually their dynamics can be considered independently because of the difference between the mass of the proton and electron ($m_e \ll M_N$): due to this difference, the electrons are faster with respect to the nuclei and can instantaneously adapt their wavefunction to the current nuclear positions. If the electron system is much "faster" than ionic one, the nuclei can be treated "classically" and the nuclear kinetic energy can be neglected (or treated as a small perturbation, if necessary).

To implement this approximation one can expand the wave function $\Psi_a(\mathbf{r}, \mathbf{R})$ into a complete orthonormal set $\Psi_\varsigma(\mathbf{r}, \mathbf{R})$ of the electronic wave functions of the whole system as follows:

$$\Psi_a(\mathbf{r}, \mathbf{R}) = \sum_{\varsigma} X_\varsigma(\mathbf{R}) \psi_\varsigma(\mathbf{r}, \mathbf{R}) \tag{2.8}$$

where $X_\varsigma(\mathbf{R})$ is a function of the nuclear coordinates only. By replacing the new form of the wave function, Eq. (2.8), in Eq. (2.1), and neglecting some electron-nuclei matrix elements arising from the nuclei kinetic energy, it is possible to write two equations: one for the electronic wave function $\Psi_\varsigma(\mathbf{r}, \mathbf{R})$,

$$(\hat{T}_e(\boldsymbol{r}) + \hat{U}_{e,e}(\boldsymbol{r}) + \hat{U}_{n,e}(\boldsymbol{r}, \boldsymbol{R}))\Psi_\zeta(\boldsymbol{r}, \boldsymbol{R}) = \varepsilon_\zeta(\boldsymbol{R})\Psi_\zeta(\boldsymbol{r}, \boldsymbol{R}) \tag{2.9}$$

and another for the nuclear wave function, $X_\zeta(\mathbf{R})$, which is an eigenstate of the following Hamiltonian:

$$\hat{H}(\boldsymbol{r}, \boldsymbol{R}) = \hat{T}_n(\boldsymbol{R}) + \hat{U}_{n,n}(\boldsymbol{R}) + \varepsilon_\zeta(\boldsymbol{R}) \tag{2.10}$$

Note that the wave function $\Psi_\zeta(\boldsymbol{r}, \boldsymbol{R})$ and the eigenvalue $\varepsilon_\zeta(\boldsymbol{R})$ depend parametrically on the nuclear coordinates $\boldsymbol{R}$. In practice, in many cases the nuclei equation is not solved, and nuclei are considered classically.

Unfortunately, solving Eq. (2.9) is still a complicated task and requires more approximations. Several methods are used to solve the Schrödinger equation approximately. In this chapter only the Hartree-Fock (HF) and the density functional theory (DFT) methods are considered.

2.2 The Hartree-Fock Method

The Hartree-Fock (HF) method allows solving the Schrödinger equation approximately. The starting point of this theory is the Pauli exclusion principle. From quantum mechanics it is known that the wavefunction depends on the position and spin of each electron and that the electrons are indistinguishable and the many-electron wavefunction is antisymmetric. That means that if two electrons switch their positions the wavefunction should change its sign. The probability density, defined as $\int d\mathbf{r}|\Psi_\zeta(\boldsymbol{r})|^2$, for both cases will be identical. Also the normalisation condition should be respected: the $\int d\mathbf{r}|\Psi_\zeta(\boldsymbol{r})|^2$ must be equal to unity. The simplest wavefunction which obeys these can be constructed as follows [1]. Each electron is given a one electron wavefunction $\psi_i(x)$, where $x = (\boldsymbol{y}, \sigma)$ contains both spatial and spin electronic coordinates:

$$\psi_i(x) = R_i(\mathbf{y})S_i(\sigma) \tag{2.11}$$

which is composed of a spatial part $R_i(\mathbf{y})$ and a spin part $S_i(\sigma)$, where σ has only two possible values α or β (spin up or spin down). In general for a system composed of N_e electrons the ground state wave function (the index ζ from here on will be omitted) may be given by the product of orbitals of each electron:

$$\Psi(x_1, x_2, \ldots, x_{N_e}) = \psi_i(x_1)\psi_2(x_2), \ldots, \psi_{N_e}(x_{N_e}) \tag{2.12}$$

However, this would be incorrect because the electrons are indistinguishable and the Pauli exclusion principle would not be respected.

If we take into account these principles, the simple product of the electron functions must be replaced by a Slater determinant [1]:

$$\Psi(x_1, x_2, \ldots, x_n) = \frac{1}{\sqrt{N_e!}} \begin{vmatrix} \psi_1(x_1) & \psi_2(x_1) & \ldots & \psi_{N_e}(x_1) \\ \psi_1(x_2) & \psi_2(x_2) & \ldots & \ldots \\ \ldots & \ldots & \ldots & \ldots \\ \psi_1(x_{N_e}) & \ldots & \ldots & \psi_{N_e}(x_{N_e}) \end{vmatrix} \tag{2.13}$$

In this approximation the one electron wavefunction is described by the Slater determinant, that gives the correct description because it is antisymmetric.

The Schrödinger equation, as defined in Eq. (2.1), is solved applying the variational principle: we have to minimise the energy and find the one-electron wavefunctions. The energy is:

$$\epsilon = \int \ldots \int \Psi^* \hat{H} \Psi dx_1 dx_2 \ldots dx_{N_e} = \langle \Psi | \hat{H} | \Psi \rangle \tag{2.14}$$

where in the right hand side we introduced the Dirac bracket notation. Therefore, the total energy, ϵ, is expressed via the one-electron orbitals $\psi_i(x)$. Varying the total energy with respect to these functions, we obtain the Hartree-Fock (HF) equations for them:

$$\hat{F}(x)\psi_i(x) = \epsilon_i \psi_i(x) \tag{2.15}$$

where $\hat{F} = \hat{h} + \hat{J} - \hat{K}$ is the Fock operator with components [1]:

$$\hat{h}(x)\psi(x) = \left[-\frac{1}{2}\nabla_r^2 - \sum_{k=1}^{N_a} \frac{Z_k e}{|\mathbf{R}_k - \mathbf{r}|} \right] \psi(x) \tag{2.16}$$

$$\hat{J}(x)\psi(x) = \left[e^2 \int \frac{\rho_{HF}(x'|x')dx'}{|\mathbf{r} - \mathbf{r}'|} \right] \psi(x) \tag{2.17}$$

$$\hat{K}(x)\psi(x) = \left[e^2 \int \frac{\rho_{HF}(x'|x)\hat{P}_{xx'}dx'}{|\mathbf{r} - \mathbf{r}'|} \right] \psi(x) \tag{2.18}$$

where $\hat{J}(x)$ and $\hat{K}(x)$ are called Coulomb and exchange operators, respectively, and $\hat{P}_{xx'}$ is the permutation operator which changes x to x' in the function it acts upon. Finally,

$$\rho_{HF}(x|x') = \sum_{i=1}^{N_e} \psi(x)\psi(x') \tag{2.19}$$

is the one-electron density matrix.

To improve the HF method, it is necessary to take into account the difference between the exact and the HF energy, that is defined as the correlation energy. Higher level QC calculations are based on Moller-Plesset Perturbation theory (MP2) or Configuration Iteration method (CI). The latter is considered as exact as this method accounts for the correlation energy exactly. These methods, however, are not considered here because, due to their very high computational cost, they can only be applied to small systems, under fifty atoms. For more detailed information about these methods, the reader is referred to [1].

2.3 Density Functional Theory

Density functional theory (DFT) [2] allows one to study the properties of complex systems of atoms by reducing the quantum mechanical problem for a many-body interacting system to an equivalent problem of non-interacting particles. This is achieved using as a fundamental variable the electronic density, instead of the many body wavefunction. Nowadays DFT is a well established and popular framework to study properties of many-body systems without using empirical parameters: hence it is considered as an ab-initio method.

The theory is advantageous for several reasons: the density is an observable that can be experimentally measured (e.g. X-ray diffraction) and the dimensionality is reduced from $4N_e$ (3 coordinate for each electron and a spin) to 4 (the coordinates x, y, z and a spin).

The theoretical basis of DFT are in the Hohenberg and Kohn theorems [2] that originally referred to an electronic system subjected to an external potential.

The first Hohenberg and Kohn theorem [2] can be formulated as follows:

- it exists a one-to-one mapping of the set of "densities" $\rho(\boldsymbol{r})$ onto the "potential" U_e
- all observable are functionals of these densities, and

Furthermore, the total energy can be written as:

$$E_U^{el}[\rho] = F[\rho] + \int U_e(\mathbf{r})\rho_e(\mathbf{r})d\mathbf{r} \tag{2.20}$$

where $F[\rho]$ is the universal functional independent of the external potential $U_e(\mathbf{r})$;

The second Hohenberg and Kohn theorem permits to determine the ground state of electronic systems by minimizing the energy with respect to the charge density, by means of the variational method.

$$F[\rho] = \min_{\Psi \to \rho} \langle \Psi | (T_e + U_{ee}) | \Psi \rangle \tag{2.21}$$

The Hohenberg and Kohn theorem, which is true for each fixed electron-electron interaction $\hat{U}_{ee}$ can be used to construct an auxiliary "non-interaction" system, the Kohn-Sham (KS) system, with external potential leading to the same set of density as the original interacting system.

$$\left[-\frac{\nabla^2}{2} + v_{KS}(\mathbf{r}) \right] \psi_i^{KS}(\mathbf{r}) = \epsilon_i \psi_i^{KS}(\mathbf{r}) \tag{2.22}$$

The ground state density is:

$$\rho_e(\mathbf{r}) = \sum_i^{occ} |\psi_i^{KS}(\mathbf{r})|^2 \tag{2.23}$$

where the sum is over all occupied states. Equation (2.22) is known as the Kohn and Sham equation [3]. The term v_{KS} is:

$$v_{KS}(\mathbf{r}) = U_{\rho.e}(\mathbf{r}) + \int \frac{\rho(\mathbf{r}_1)}{|\mathbf{r}_1 - \mathbf{r}|} d\mathbf{r}_1 + v_{xc}(\mathbf{r}, \rho) \tag{2.24}$$

where v_{xc} is the exchange and correlation potential, which is an unknown functional of the electron density. The Kohn and Sham equations are solved in a self-consistent way with respect to the orbitals $\psi_i^{KS}(\mathbf{r})$ entering the charge density via Eq. (2.23).

The total energy of the electronic system has the following form:

$$E^{el}[\rho] = -\frac{1}{2}\sum_{i=1}^{occ} \int \psi_i^* \nabla^2 \psi_i d\mathbf{r} + \frac{1}{2}\int \frac{[\rho(\mathbf{r})\rho(\mathbf{r}')]}{|\mathbf{r} - \mathbf{r}'|} d\mathbf{r}d\mathbf{r}' + \int U_e(\mathbf{r})\rho(\mathbf{r})d\mathbf{r} + E_{xc}[\rho] \tag{2.25}$$

Here $E_{xc}[\rho]$ represents the exchange and correlation energy. To solve the Kohn and Sham equations we need to determine the form of $E_{xc}[\rho]$ (or v_{xc}).

An approximation introduced by Kohn and Sham to solve the problem is known as the *Local Density Approximation* (LDA). It assumes that the exact exchange and correlation potential at the point **r** is equal to the exchange correlation energy of a homogeneous electron gas, which has the same density as in the real system at point **r**. In this approximation the density is purely local, the corrections to the exchange and correlation energy at the point **r** due to nearby inhomogeneities in the electron density are ignored.

Another popular approximation is the *Generalized Gradient Approximation* (GGA), in which the correlation energy also depends on the density gradient. It takes account of the density inhomogeneities.

Both methods are very popular in practical calculations. One of the failures of the standard approximations is the incorrect description of weak vdW forces. The dispersion (van der Waals) interaction is inherently non-local, so that local (LDA) and semi-local (GGA) functionals fail to describe correctly interactions between two systems at distances from each other where their electronic densities do not overlap.

2.4 Technical Details of DFT Calculations

There are few practical issues to be faced with when one uses the DFT method. These concern the choice of the basis set, k-point sampling used for the description of the electronic wavefunctions, the pseudopotential and the periodic boundary conditions (PBC).

2.4.1 Choice of the Basis Set

To solve the KS equation it is necessary to expand the electronic wavefunction in a basis set and there is a quite wide variety of basis sets used for constructing the electronic wavefunction. This can be obtained using a sum over atomic orbitals [1]:

$$\psi_\lambda(\mathbf{r}) = \sum_b C_{b\lambda}\phi_b(\mathbf{r}) \tag{2.26}$$

where $C_{b\lambda}$ are the expansion coefficients. These basis functions are usually constructed from atomic orbitals centred on atoms [1]:

$$\phi_b(\mathbf{r}) \rightarrow \xi_{nlm}(\mathbf{r} - \mathbf{R}_A) \tag{2.27}$$

where $\mathbf{R}_A$ is a vector of an atom A and n, l and m are the quantum numbers. The basis function ξ_{nlm} can be split into two parts, one radial and another angular, i.e. $\xi_{nlm} = R_{nl}(\mathbf{r})S_{lm}(\mathbf{r})$ [1]. The angular part, S_{lm}, is a spherical harmonic describing the angular dependence while the radial part, $R_{nl}(\mathbf{r})$, describes the radial dependence. There are three main types for the orbitals which can be obtained following these principles [1]:

- Slater type orbitals where the radial part of the atomic orbital is a simple exponential which mimics the exact eigenfunctions of the hydrogen atom.
- Gaussian orbitals which are similar to the Slater type orbitals but the exponential terms in the radial functions are replaced by Gaussian functions.
- Numerical orbitals which have radial functions specified on a numerical grid which are more suitable for quick calculations.

The size of the basis set is important for the calculations: the larger the basis set, the more accurate the calculation should be. In this approximation the basis set suffers from being incomplete, with practical implications for the results, especially binding (or stabilisation) energies. Sometimes, adding more atomic orbitals may result in over-completeness problem as well, so one has to be very careful in reaching convergence with this type of basis set.

In fact, there are other ways to create the basis set where periodicity is present. It is possible to split the basis set into two main types [1]:

- Crystalline atomic orbitals where the atomic orbitals constructed using either of the atom based approaches given above are periodically repeated . The SIESTA code (see Sect. 2.5) uses crystalline numerical atomic orbitals.
- Plane waves [1]:

$$\phi_{\mathbf{k}+\mathbf{G}}(\mathbf{r}) = \frac{1}{\sqrt{V}} e^{i(\mathbf{k}+\mathbf{G})\mathbf{r}} \tag{2.28}$$

where **k** is a vector in the Brillouin zone, **G** is a reciprocal lattice vector and V is the volume of the crystal. These are not atom centred functions and they are completely analytic. They are also easy to use with periodic boundary conditions (PBC), especially for solids. However, with empty space in the system (e.g. a molecule on a surface in the slab model) they are very inefficient because the number of plane waves needed becomes very large.

2.4.2 Pseudopotential Method

Another important tool to solve the KS equation is the pseudopotentials method [1]. In general the number of electrons in a system is determined by summing up all core and valence electrons, where the latter are known to be most important for chemical bonding. This is due to their wavefunctions being further away from the nuclei and this also has the effect that they are shielded from nuclei Coulomb attraction by core electrons. The core electrons feel a very strong Coulomb attraction from the nuclei so that their wavefunctions are very localised around it and less affected by the surrounding of the atom.

In fact, the core electrons can be considered as frozen, i.e. in a molecular or periodic system they can be considered as having the same form as they have in the isolated atom. Therefore, only the valence electrons need to be considered variationally. However, valence electrons need to be orthogonal to the core electrons and therefore will oscillate rapidly in the core region. This requires many basis functions being required to properly describe the valence wavefunctions in the core regions [1].

This is where the introduction of pseudo-wavefunctions for the valence electrons becomes necessary. These pseudo-wavefunctions have unnaturally smooth behaviour inside the core regions, but still keep their natural form outside the core regions so that the chemistry is unaffected. Pseudopotentials incorporate the potential due to the nuclei and the core electrons into one effective potential. There are three general forms for the pseudo-potential [1]:

- Norm-conserving pseudopotentials which provide the correct normalisation for the pseudo-wavefunctions.
- Ultra-soft pseudopotentials which do not provide the correct normalisation for the wavefunctions, but are smoother in the core region. This complicates the KS equation but improves the efficiency overall.
- The Projector Augmented Wave method which takes explicit account of the frozen core electrons in a way which still provides good efficiency.

The radius of the core region is denoted by r_c. One must be careful in choosing r_c: if it is too large, the chemistry in the system may be unreliable; if it is too small, the oscillating characteristics in the core region need to be described. This would destroy the reason for introducing the pseudopotential as larger basis sets would be needed.

2.4.3 Periodic Boundary Conditions and k-Point Sampling

Many physical systems have translational symmetry (for instance, solids, monolayers, chains etc.). However, even if this is not the case (e.g. isolated molecules, a point defect in a solid, an adsorbed molecule on a surface), periodicity in all the three dimensions is imposed artificially. When one system has only one or two-dimensional periodicity, the 3D periodicity can be easily implemented by introducing large vacuum gaps between the periodic systems so that one or two dimensional structures cannot spuriously interact with their periodic images.

The electronic structure of a system with PBC resembles a band structure of a perfect periodic crystal [1]. In this case the KS potential (2.24) is translationally invariant

$$v_{KS}(\boldsymbol{r}) = v_{KS}(\boldsymbol{r} + \boldsymbol{R}) \tag{2.29}$$

where $\boldsymbol{R}$ is the direct lattice vector. According to the Bloch theorem, this property will affect the form of the single-particle wavefunctions. One electron energies ε_λ and the wavefunction ψ_λ are counted using the band index n and the k point within the Brillouin zone, i.e. via λ which is the joint index ($n\mathbf{k}$):

$$\psi_\lambda(\mathbf{r}) = \psi_{n\mathbf{k}} = e^{i\mathbf{kr}} u_{n\mathbf{k}}(\mathbf{r}) \tag{2.30}$$

$$u_{n\mathbf{k}}(\mathbf{r}) = u_{n\mathbf{k}}(\mathbf{r} + \mathbf{R}) \tag{2.31}$$

In practice, the number of **k**-points to be used depends on the size of the unit cell. When the unit cell is big, the number of **k**-points required is small because the corresponding Brillouin zone to sample is smaller [1]. The actual number of **k**-points required for given system can be chosen by checking the convergence of the energy and geometry while increasing the number of **k**-points.

2.5 Siesta Method

The DFT code mainly used in this thesis for calculations of the electronic structure and geometry of various systems is the Spanish Initiative for Electronic Simulation with Thousands of Atoms (SIESTA) [4]. The code employs as the basis set a linear combination of numerical orbitals, uses periodic boundary conditions and the method of pseudopotentials. The core electron properties are simulated using scalar-relativistic norm-conserving [5, 6] Troullier-Martins [7] pseudopotentials factorised in the Kleinman-Bylander form [8]. In this work the nonlinear core-valence exchange correlation scheme [9] is applied for all elements except for hydrogen.

In all our calculations, a double-zeta-plus polarisation orbitals (DZP) basis was used. DZP uses two functions for each atomic orbital plus polarisation orbitals. The energy cutoff used is between 160–200 Ry.

Perdew, Becke and Ernzerhof (PBE) [10] generalised gradient approximation (GGA) functional is used in this thesis for the exchange and correlation energy. This functional gives good description of the hydrogen bond formation of DNA pairs [11–14] as compared to quantum chemistry calculations.

The atomic relaxation in our systems was performed until the forces on atoms were no larger than 0.01 eV/Å for all the individual molecules and pair calculations. The exception were the monolayers and the molecules on the gold surface, where the precision was of 0.05 eV/Å or better.

In the calculations on single molecules and dimers, very large super-cells were used to avoid unphysical interaction between the images.

2.5.1 Definition of Useful Energies to Analyse Systems Stability

In order to analyse the stability of the studied systems, it is necessary to calculate a number of useful energies [15]. The first of these is the stabilisation energy, E_{stab}, defined as the total energy of the relaxed combined system (i.e. a pair of molecules) minus the total energies of all its individual components (two molecules) relaxed separately. The combined system is stable if the E_{stab} is negative. To describe the interaction between the single molecules, we define the interaction energy, E_{int}, as the total energy of the system minus the energies of each individual molecule calculated in the geometry of the relaxed combined system. This energy should also be negative for a stable system. Finally, we define the deformation energy, $E_{\mathrm{def}} = E_{\mathrm{stab}} - E_{\mathrm{int}}$, which characterises the sum of losses in the total energies of the molecules due to their relaxation towards their geometry in the combined system. The E_{def} is positive because the energies of the individual molecules calculated in the geometry of the combined system are higher than when they are individually relaxed.

These energies are correct only if the complete basis set is used. Therefore, if a finite localised basis set is used, a Basis Set Superposition Error (BSSE) [16] correction is to be applied to calculate binding energies. Moreover, since we employ a localised basis set, these energies would only be correct after the application of the BSSE correction. Therefore, to simplify the analysis, the same BSSE correction was applied to the interaction and stabilisation energies ensuring that this exact relationship remains. These corrections have been calculated by the standard Boys-Bernardi counter poise correction method [16]. The BSSE correction for a pair is defined as:

$$E_{\mathrm{BSSE}} = E_A + E_B - E_A^* - E_B^* \tag{2.32}$$

where E_A and E_B are the total energies of each monomer in the geometry of the pair, and E_A^* and E_B^* are the total energies of each molecule in the geometry of the pair with the orbitals of the other molecule also present. For this second group of energies the basis set is larger as more orbitals are available in the construction, implying a lower energy as the basis set is more complete (e.g. $E_A^* < E_A$), then the E_{BSSE} is always positive. In the calculation the additional orbitals are considered as "ghost" orbitals because they are associated with atoms that do not exist in calculations of E_A^* and E_B^*.

2.5.2 Characterisation of the Hydrogen Bonds: "Kebab" Structures

A way of characterising the strength of the hydrogen bonding between two molecules is by analysing the electron density difference plots [11–14], $\Delta\rho = \rho_{ab} - (\rho_a + \rho_b)$, where ρ_{ab} is the electron density of the combined pair and ρ_a and ρ_b are the densities of the individual monomers calculated at the geometry of the pair. Hydrogen bonds are characterised by a "kebab" structure of alternating regions of charge excess and depletion, that represent the redistribution of the charge density due to the hydrogen bond formation Kelly et al. [11]. The stronger the bond, the more regular is the "kebab" structure. This concept can be easily generalised for complexes containing more than two molecules.

2.6 Diffusion Calculations

To investigate the corrugation of the surface potential for a single molecule or a cluster of molecules and hence calculate the corresponding diffusion barriers, we placed the molecules above the gold surface (the DFT optimum height) and translated it in steps of 0.1 Å along several directions across the surface ($[\bar{1}10]$, $[10\bar{1}]$ and $[11\bar{2}]$); during each relaxation the lateral position of one atom of the molecules was fixed, although it was allowed to move vertically. All other atoms of the system including the top layer atoms of the surface were allowed to relax. The bottom atoms of the surface slab were fixed in the bulk-line position.

2.7 Van der Waals Implementation in DFT Method

The standard approximations to the exchange-correlation energy in DFT, discussed in Sect. 2.3, are relatively cheap and can describe reliably many characteristics of systems of interest. However, LDA and GGA approximations to DFT are not able to describe a weak non-local dispersion interaction, also known as the as the vdW interaction, which is essential for many systems. The reason why the standard DFT functionals fail is due to an incorrect description of the electron correlation in situations where the electron densities of two parts of the system do not overlap and the vdW interaction becomes dominant. Indeed, the vdW interaction is non-local in character and is entirely missing in most of the GGA density functionals which are semi-local.

There have been attempts to solve the problem in several ways. One approach consists of adding to the standard DFT method additional semiempirical terms which take into account the dispersion interaction (DFT-D method) [17]. However, this method has a limited flexibility and is system dependent. The parameters are bound to work for the systems they were designed for and the use of these parameters outside that set of parametrisation may not be able to deliver reasonable results. Another approach includes accounting for the vdW interaction through effective atom-electron pseudopotentials [18]. Recently, a truly first-principles approach, which does not require any fitting parameters and is within the remit of the DFT method was developed (vdW-DF) [19–23] and successfully applied to many systems. In this method the exchange-correlation energy functional has the form:

$$E_{vdw-DF}[\rho] = E_{\mathrm{GGA}}[\rho] - E_{\mathrm{GGA,c}}[\rho] + E_{\mathrm{LDA,c}}[\rho] + E_c^{nl}[\rho] \tag{2.33}$$

where ρ is the electron density, E_{GGA} is the GGA exchange-correlation energy corresponding to some flavour of GGA and $E_{\mathrm{LDA,c}}$ is the LDA correlation energy. The E_c^{nl} represents a non-local contribution to the correlation energy and accounts for the dispersion interaction. The difference of correlation energies, $E_{\mathrm{LDA,c}}[\rho] - E_{\mathrm{GGA,c}}[\rho]$, serves to replace the correlation energy contained in GGA functional with that of the LDA. This particular form of the energy expression guarantees that for uniform densities (e.g. the uniform electron gas) the correlation energy equals exactly that of the LDA density functional as the non-local term $E_c^{nl}[\rho]$ vanishes.

This approach has been implemented in SIESTA [24, 25] and used in some calculation [24, 26]. It was shown that inclusion of dispersion interaction may result in substantial changes of the stabilisation energies of molecules on surfaces.

2.8 Super-Structure Modelling: A Systematic Method

The STM and AFM techniques which allow the imaging of the surface covered with molecules was mentioned in Sect. 1.3. Unfortunately, it is not possible to achieve atomic resolution with any of these techniques for most of the organic molecules. Therefore, a theoretical approach is required to determine the geometries of the observed supramolecular assemblies.

This subsection describes the approach employed to construct the supramolecular structures and the computational method used. In order to build all the possible structures, a systematic approach has been used [11], which consists of the following steps: (i) identification of all the peripheral sites of the molecule that can participate in a hydrogen bonding with another molecule; (ii) all possible dimers are constructed; (iii) by combining molecules, using dimer rules, all unit cells are constructed having a predefined number of molecules; (iv) all possible one dimensional (1D) structures (chains) are constructed for every unit cell; (v) by attaching chains parallel to each other, all possible two dimensional (2D) structures are formed; (vi) the stabilities per unit cell of the assemblies predicted in this way are estimated by summing up all the dimer energies per cell; (vii) the most stable predicted structures are then fully relaxed using DFT. In this way, by assuming a certain number of molecules per unit cell, one can build up all the possible 1D and 2D structures. The gas phase calculation (i.e. without taking into account the surface) is a good approximation by considering that the molecules are quite mobile on the gold surface. This is because the surface potential, as will be demonstrated in Sect. 4.1.1 is flat. This approach is particularly useful because it allows considering all possible structures in the gas phase with a modest computational cost.

2.9 Modelling of STM Images

2.9.1 Tersoff-Hamann Method

The modelling of STM images could be a useful tool to gain an understanding of experimental images and as a means of identifying the actual system geometry.

The Tersoff-Hamann method provides an easy way to analyse this surface phenomenon.

According to the Fermi's golden rule one can write the current flowing from the tip to the sample using the Bardeen's expression [27]:

$$I_{t\rightarrow s} = \frac{2\pi e}{\hbar}\int |M_{ts}|^2 N_t(E-eV)N_s(E)f_t(E-eV)[1-f_s(E)]dE \qquad (2.34)$$

$f(E)$ denotes the Fermi-Dirac distribution for the tip and the sample, $N_t(E)$ and $N_s(E)$ denote their density of states. M_{ts} is the effective matrix element for the current between the tip and sample due to the states of energy E, and V is the applied voltage.

In order to model the STM image of the molecules on the surface the Tersoff and Hamann model has been used [28]. Tersoff and Hamann suggested replacing the unknown electronic structure of the tip by a simple model, where the wave function of the tip atom is assumed to be an atomic s-wave function. Since the tunneling current depends on the overlap of the wavefunction of the tip and sample, and the wavefunction decays exponentially into the vacuum, only the orbital localised at the tip will be involved in the tunneling matrix element.

Within the approximation of zero temperature and low bias voltage the total current flowing from the tip to the sample will be:

$$I = \frac{2\pi e^2}{\hbar} V \sum_N |M_{ts}|^2 \delta(E_s - E_F)\delta(E_t - E_F) \tag{2.35}$$

where V represents the voltage applied and δ-functions ensure the energy conservation.

The effective matrix element M_{ts} is

$$M_{ts} = \frac{\hbar^2}{2m} \int (\Psi_t^* \nabla \Psi_s - \Psi_s \nabla \Psi_t^*) dS \tag{2.36}$$

where the integral is taken over a surface separating the tip and sample. To calculate the matrix element, the surface wave-function Ψ_s is written as a 2D reciprocal lattice expansion.

The main point of Tersoff-Hamann theory is that no choice of the wavefunction of the tip is needed. Since there is usually no knowledge of the exact tip structure, Ψ_t can be modelled as a solution in a locally spherical potential with curvature R about its center r. So asymptotically the tip wavefunction is chosen to have the form of an s-wave. By using this approximation the matrix element becomes proportional to the sample wavefunction evaluated at the tip center of curvature ($M_{ts} \propto \Psi_s(r_0)$). Then the total current flowing from Eq. (2.35) can be written as:

$$I \propto V N_t(E_F) \sum_s |\Psi_s(r_0)|^2 \delta(E_s - E_F) \tag{2.37}$$

In this expression, the sum represents the local density of states of the sample (LDOS) at the Fermi level evaluated at the tip center. Then Eq. (2.37) can be rewritten as

$$I \propto V N_t(E_F) \mathrm{LDOS}(r_0, E_F) \tag{2.38}$$

This equation gives an easy way for the interpretation of the STM images.

SIESTA can output the LDOS integrated over a range of energies which allows one the calculation of simple STM images in the Tersoff-Hamann approximation.

2.9.2 Scattering Formalisms

The Tersoff-Hamann approximation is an easy way to analyse the surface phenomenon. However, this kind of approach does not take into account the effect of the tip.

We also used a new program called Nt_STM which is a graphical interface for "Green", a code written by Cerda [29, 30]. This approach is based on the generalised Landauer theory [31], and the scattering matrix is determined by the Green's-function techniques [29].

The junction is modelled by partitioning the system into three regions: two bulk regions for the substrate and the tip, and two principal layers that are involved in the scattering. The bulk region for the tip and the substrate is periodically repeated in the z direction away from the junction while the apexes involved in the scattering are composed by several atomic layers. The whole system is considered periodic in the x and y directions.

The tunnelling process is considered as a scattering mechanism for which the scattering matrix S is calculated from the electronic transmission and reflection amplitudes at the tunnel junction. The conductance is evaluated via the Landauer-B$\ddot{u}$ttiker formula for the multichannel case[1] [31].

At the zero temperature limit, the current I is

$$I(V) = \frac{e}{\pi\hbar} \int_{0}^{-eV} T^{\pm}(E + \mu_T) dE \tag{2.39}$$

The $T^{\pm}(E)$ is the total transmission coefficient:

$$T^{\pm}(E) = \sum_{m,m'} T^{\pm}_{mm'}(E) \tag{2.40}$$

where $T^{\pm}_{mm'}(E)$ are the matrix elements that give the probability for an electron in the m channel to be transmitted to channel m'. The electron propagates from the channel m at the negative side of the tunnel junction to the channel m' at the positive side without dissipation. Each channel wavefunction has the current bloch-line asymptotic behaviour in the regions far away from the junction.

2.10 Classical Molecular Dynamics

Classical molecular dynamics uses empirical potentials (known as the "force field") that either ignore quantum mechanical effects, or attempt to capture them in a limited way through entirely empirical equations. Parameters in the potential are fitted against known physical properties simulated in the system, such as the elastic constants and the lattice parameters, in order to reproduce the basic aspects of the intermolecular and intramolecular interactions.

In order to perform classical molecular dynamics calculations, the Sci-Fi code has been used [32]. The energy expression implemented in the code has the form:

$$U = V_{\text{cov}} + V_{vdW} + V_{el} \tag{2.41}$$

[1] At a given energy E the surface (s) and the tip (t) could be described by a certain number $N_s(E)$ or $N_t(E)$ of independent conductors that are defined as channels.

where in that energy, the effects of the covalent (V_{cov}), van der Waals (V_{vdW}) and the electrostatic (V_{el}) interactions are taken into account. The covalent term is used for describing internal energies of organic molecules and it consists of a number of terms:

$$V_{\text{cov}} = \sum_{b}^{Nb} k_b(b-b_0)^2 + \sum_{\theta}^{N_\theta} k_\theta(\theta-\theta_0)^2 + \sum_{\phi}^{N_\phi} K_\phi[1+\cos(n\phi-\phi_0] + \sum_{\chi}^{N_\chi} K_\chi[1+\cos(n\chi-\chi_0)] \quad (2.42)$$

The terms in the above expression apply to the many body interactions in the system and summations only apply to specified interactions, in this case only covalent interactions. These interactions are most conveniently specified via the internal coordinates of the system, not via the Cartesian coordinates of atoms as is the case for other interactions. These internal coordinates are: bond lengths b, angles between bonds ϑ, torsion angles about bonds φ and out-of-plane angles χ. The vdW energy is given via

$$V_{vdW} = \sum_{i<j}^{\text{atoms}} \left(\frac{A_{ij}}{R_{ij}^{12}} - \frac{B_{ij}}{R_{ij}^{6}} \right) \quad (2.43)$$

A_{ij} and B_{ij} are constants and R_{ij} is the distance between atoms i and j. It also includes short-range repulsion terms.

Finally, the electrostatic energy V_{el}, in Eq. 2.41, describes electrostatic interaction between atoms.

Here, in a number of calculations we combined a DFT with a force field [33] developed by fitting results of the quantum chemistry calculations at the MP2 level of various groups of all DNA base molecules with gold clusters. Note that the force field includes all interactions between the bases and the gold surface, including vdW interactions In these hybrid calculations only interaction between molecules and surfaces was supplemented by a force field. We have also done some calculations using the Sci-Fi code [32] we did not relax neither the surface, nor the melamine molecule, so that the total energy of the melamine-surface system represents directly the adsorption energy.

The following chapters review the results obtained. In Chap. 3, attention is paid to the assembly of molecules in the gas phase, considering the molecule-molecule interaction and disregarding the presence of the surface. Chapters 4, 5 include a more detailed characterisation of the interaction between molecules and surface.

References

1. Kantorovich LN (2004) Quantum theory of the solid state: an introduction. Fundamental theories of physics. Kluwer Academic Publishers, Dordrecht
2. Hohenberg P, Kohn W (1964) Phys Rev Lett 136:864B
3. Kohn W, Sham LJ (1965) Phys Rev Lett 140:1133A
4. Sanchez-Portal D, Ordejon P, Artacho E, Soler JM (1997) Density functional method for very large systems with lcao bas is sets. Int J Quantum Chem 65:453
5. Kleinman L (1980) Phys Rev B 21:2630
6. Bachelet GB, Schluter M (1982) Phys Rev B 25:2103
7. Troullier N, Martins JL (1991) Phys Rev B 43:1993
8. Kleinman L, Bylander DM (1982) Phys Rev Lett 48:1425
9. Louie SG, Froyen S, Cohen ML (1982) Phys Rev B 26:1738
10. Perdew JP, Burke K, Ernzerhof M (1996) Phys Rev Lett 77:3865
11. Kelly REA, Lee YJ, Kantorovich LN (2005) J Phys Chem B 109:22045
12. Kelly REA, Kantorovich LN (2006) J Mater Chem 16:1894
13. Kelly REA, Lee YJ, Kantorovich LN (2005) J Phys Chem B 109:11933
14. Kelly REA, Lee YJ, Kantorovich LN (2006) J Phys Chem B 110:2249
15. Sponer J, Leszczynski J, Hobza P J Phys Chem
16. Boys F, Bernardi F (1970) Mol Phys 19:553
17. Grimme S (2004) J Comput Chem 25:1463
18. von Lilienfeld OA, Tavernelli I, Rothlisberger U, Sebastiani D Phys Rev Lett
19. Dion M, Rydberg H, Schroeder E, Langreth DC, Lundqvist BI (2004) Phys Rev Lett 92:246401
20. Langreth DC, Dion M, Rydberg H, Schroder E, Hyldgaard P, Lundqvist BI (2005) J Quantum Chem 101:599
21. Thonhauser T, Cooper VR, Li S, Puzder A, Hyldgaard P, Langreth DC (2007) Phys Rev B 76:125112
22. Cooper VR, Thonhauser T, Puzder A, Schroeder E, Lundqvist BI, Langreth DC (2008) J Am Chem Soc 130:1304
23. Langreth DC, Lundqvist BI, Chakarova-Draxl SD, Cooper VR, Dion M, Hyldgaard P, Kelkkanen A, Kleis J, Kong LZ, Li S, Moses PG, Murray E, Puzder A, Rydberg H, Schroder E, Rydberg T, Thonhauser H (2009) Cond Matter 21:084203
24. Gulans A, Puska MJ, Nienminen RN Phys Rev B
25. Roman-Perez G, Soler JM Phys Rev Lett
26. Mura M, Gulans A, Thonhauser T, Kantorovich L (2009) Role of van der waals interaction in forming molecule-metal junctions: flat organic molecules on the au(111) surface. Phys Chem Chem Phys (submitted)
27. Bardeen J (1961) Phys Rev Lett 6:57
28. Tersoff J, Hamann DR (1985) Stm Theory Phys Rev B 31:805
29. Cerdá J, Van Hove MA, Sautet P, Salmeron M (1997) Efficient method for the simulation of stm images. i. generalized green-function formalism. Phys Rev B 56(24):15885–15899
30. Cerdá J, Van Hove MA, Sautet P, Salmeron M (1997) Efficient method for the simulation of stm images. ii. application to clean rh(111) and rh(111)+c(4x2)-2s. Phys Rev B 56(24):15900–15918
31. Büttiker M, Imry Y, Landauer R, Pinhas S (May 1985) Generalized many-channel conductance formula with application to small rings. Phys Rev B 31(10):6207–6215
32. Kantorovich LN, Trevethan T, Polesel-Maris J, Foster A Self consistent image dorce interaction + virtual AFM machine
33. Piana S, Bilic A (2006) J Phys Chem B 110:23467

Chapter 3
Hydrogen-Bonding Templates in the Gas Phase

In this chapter we shall consider all possible gas-phase structures which could be formed by various molecules on the Au(111) surface. To understand these findings, we followed a systematic theoretical approach in constructing 1D and 2D planar periodic molecular assemblies. The method is described above, see Sect. 2.8. Our method is based on predicting all possible structures in the gas phase by first identifying and then utilising all connections between the molecules. All possible monolayers based on dimers or more complicated unit cells could be obtained using this method. This will be done firstly in the gas phase, that is justified if the potential energy surface (PES) of the "molecules + surface" system is flat. This implies that the molecules diffuse easily on the surface and the assembly is driven mainly by molecule-molecule interaction.

In the next section we shall apply the systematic approach to different molecules such as Melamine, PTCDA, PTCDI. NTCDA, NTCDI and mixed phases of melamine and PTCDA/PTCDI.

3.1 Melamine

We shall consider all possible gas phase structures based on up to four melamine molecules, see Fig. 3.1a, per unit cell. Initially, we shall describe in detail the approach used and present the melamine dimer; this will be followed by building all superstructures based on the dimer unit cell. After that, we shall consider unit cells containing three and four molecules and the superstructures based on these unit cells.

3.1.1 Dimer

The starting point in our method is the identification of the binding sites of the melamine molecule. We identified six sites, which, due to the high symmetry of

M. Mura, *Self-Assembly of Flat Organic Molecules on Metal Surfaces*,
Springer Theses, DOI: 10.1007/978-3-642-30325-8_3,

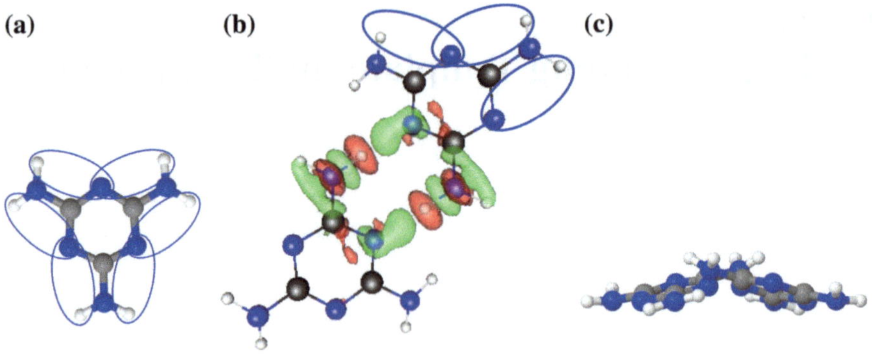

Fig. 3.1 **a** Geometry of the melamine molecule and its six binding sites shown by *ovals*; **b** planar melamine dimer shown together with the plot of the electron density difference corresponding to 0.01 electrons/Å^3. The dimer binding sites on one of the molecules are also indicated; the other molecule has similar sites. The *green* surfaces correspond to the regions of positive electron density difference (excess) and the *red* areas correspond to the regions of negative electrons density difference (depletion); **c** side view of the non-planar dimer

melamine (point group D_{3h}), are equivalent; each site is able to form a double hydrogen bond with similar sites of another melamine molecule. They are composed of a nitrogen atom, which acts as the hydrogen acceptor, and an amino group, which is the hydrogen donor of the hydrogen bond; see Fig. 3.1b.

Only a single dimer can be constructed in the case of melamine [1], shown in Fig. 3.1b. The pair is built by rotating the second melamine by 60° with respect to the first one, and connecting the second molecule to the first to form a double hydrogen bond of the N–H...N type.

Surprisingly, the melamine dimer relaxed into two stable geometries: one where the two rings are in the same plane (planar) and another one where the rings are not planar, but are at some angle to each other, as shown in Fig. 3.1c. The top view and the "kebab" structure of the electron density difference of the two dimers, shown for the planar dimer in Fig. 3.1b as an example, are practically the same. The two structures are equally stable with similar hydrogen bond properties (lengths and angles), as shown in Table 3.1. The non-planar structure is more favorable, however, the energy difference is within error bars of DFT. The two configurations are possible because the melamine molecule is not planar in the bulk [2] and in the gas phase. As found in Ref. [2] and confirmed by our calculations, the hydrogen atoms are out of the ring plane by the same distance z. The planar structure is also in agreement with previous calculations [3, 4]. Since the planar structure is more likely to be realised on the surface, in the following analysis we will focus on the planar dimer only.

The stabilisation energy, E_{stab}, per hydrogen bond in the melamine dimer is -0.22 eV for the planar case and -0.24 eV for the non-planar structure; both hydrogen bonds in either of the dimers are identical and bring equal contributions to the stabilisation of the dimer structure. The results of the dimer calculations are needed for our further analysis, because at the next step we shall consider super-structures

Table 3.1 Energies and geometrical parameters of the melamine dimers

	Planar	Non-planar	Literature
E_{stab} (eV)	−0.44	−0.48	−0.42 [5]
E_{int} (eV)	−0.57	−0.60	–
E_{def} (eV)	0.13	0.12	–
d_{N}–H...N (Å)	2.89	2.89	2.9 [6, 7]
N–$\widehat{H}$...N (°)	178.6	177.6	–

The stabilisation and interaction energies include the BSSE correction. The last two rows show the distance between the N atom involved in the N–H bond in one molecule and the N atom in the other molecule and the N–H...N angle. Available results from other calculations are given in the last column

Table 3.2 Comparison of melamine dimer binding energies (in eV) obtained using PBE, MP2 and vdW-DF based calculations

	PBE	vdW-DF	MP2
M–M	−0.48 [11]	−0.42	−0.49

based on the dimer. Also note that the stabilisation energies of complex periodic structures based on melamine dimers can be estimated as the stabilisation energy per bond multiplied by the number of hydrogen bonds per unit cell in the periodic structure.

3.1.1.1 The Effect of Dispersion Force on the Melamine Dimer Formation

In order to evaluate the influence of the dispersion forces on the stabilisation energy of the melamine dimer, we also performed its relaxation with the vdW-DF method. In order to evaluate the performance of the two density functionals, we also compare our binding energies of the dimers with that obtained using the MP2 method by Gulans.[1] As demonstrated by Jurečka et al. [8], it reliably describes hydrogen-bonded interactions. On the other hand, the benchmark calculations of Goll et al. [9] imply a good performance of the PBE functional for hydrogen-bonded complexes. The result of the binding energy for the hydrogen bonded melamine dimer is presented in Table 3.2. In the case of the strongly bound melamine dimer, we have also looked at charge density differences and found a characteristic for the hydrogen bonding "kebab" structure, which is very similar to that obtained with the PBE density functional results of Mura et al. [10].

We can conclude from these calculations that the vdW-DF method gives a description of the lateral interaction between molecules which is very close to that provided by GGA functionals such as PBE.

[1] Department of Applied Physics, COMP, Helsinki University of technology, P. O. Box 1100, 02015 TKK, Finland.

3.1.1.2 One-Dimensional Chains Based on the Melamine Dimer

In this section we will present the 1D structures that can be built using the melamine dimer as a unit cell. As before, we start from the identification of the binding sites available on the periphery of the dimer: there are three non similar sites in the dimer, see Fig. 3.1b, which are available for the connection with more dimers. We build 1D chains by connecting dimers in a chain-like manner along one direction. The chains are obtained by adding a second dimer to a chosen binding site of the first dimer, then a third dimer is added to the same site of the second one, and so on. In this way, and using all three dimer binding sites, three different chains can be constructed, see Fig. 3.2. One more chain, based on the dimer fourth binding site, appears to be equivalent to one of the three upon a rotation by 60°.

The one-dimensional structures, shown in Fig. 3.2, are all planar. The geometric properties of the hydrogen bonds in the chains are in good agreement with the values found for the melamine dimer, compare Tables 3.1 and 3.3; only for Chain3 the bond length is somewhat longer. Also, the first two chains are similar in their stability and geometric parameters, whereas Chain3 is somewhat different. Its stabilisation energy, −0.70 eV, is smaller than those of the other chains and is also much less than the stabilisation energy of −0.88 eV predicted using the dimer energies (for all three chains there are four hydrogen bonds per unit cell). Note that, for the first two chains, their stabilisation energies are in good agreement with the predicted value, −0.88 eV. Chain3 also has a more compact structure, as it has the shorter lattice vector, 7.76 Å, against 10.70 and 12.26 Å of the first and the second chains. Chain3, which has a close-packed configuration, is indeed an interesting case because it is almost a two-dimensional structure by itself.

Electron density difference plots, shown in Fig. 3.2, demonstrate that in the case of Chain3 the hydrogen bonds are much weaker: one amino group out of three of each molecule participates in two hydrogen bonds at the same time, thus making it difficult for the corresponding nitrogen donor to provide enough electron density for the bonds. In addition, there is also a steric effect: some of the hydrogen atoms in the third chain are too close in distance; consequently, in order to minimise the deformation energy, the hydrogen atoms of neighbouring molecules move away from the molecular plane in opposite directions. All these factors make Chain3 less stable than the other chains, however, it is much denser and it allows a more efficient packing of the molecules.

Thus, according to our calculations, three chains are possible with Chain1 being the most stable and Chain3 being the least stable.

In the next section we will present how, starting from the 1D structures, 2D structures can be derived.

3.1.1.3 Two-Dimensional Structures Based on the Melamine Dimer

To build all possible 2D structures, we considered each chain separately and again identified the binding sites on either side of the chain which are available to connect

(a)

(b)

(c)

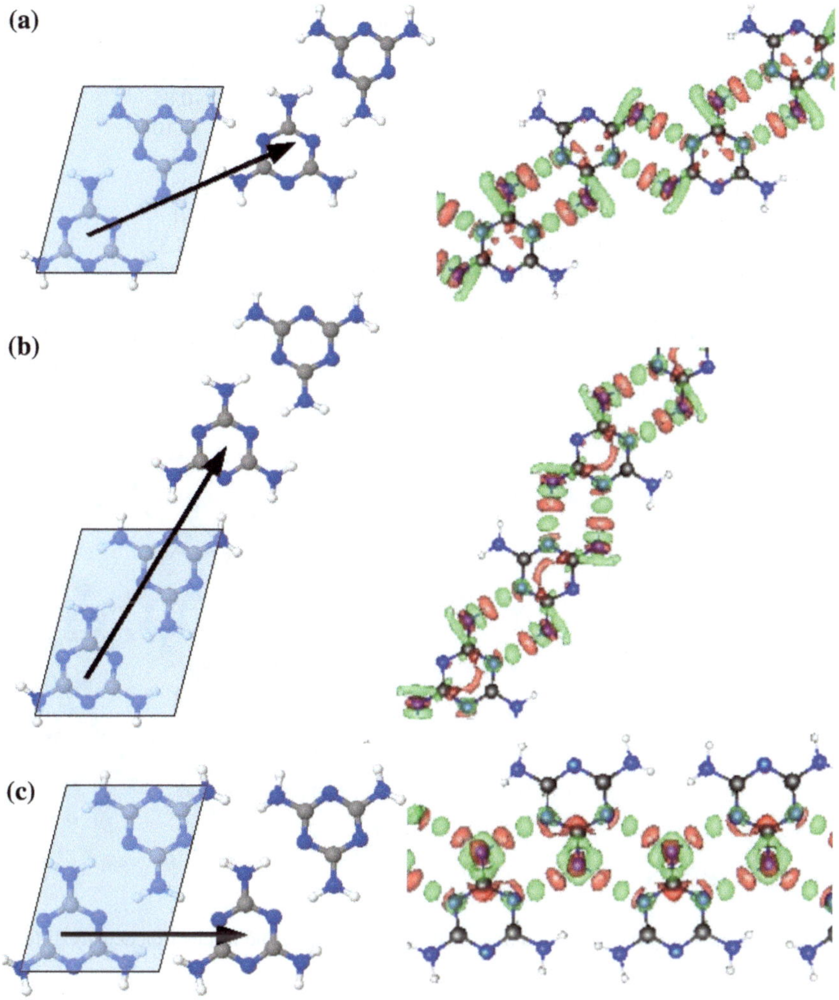

Fig. 3.2 The geometries (the *left column*) and the electron density difference plots (the *right column*, see the caption to Fig. 3.1 for notations) for the three melamine chains: **a** Chain1; **b** Chain2 and **c** Chain3. The unit cell and the lattice vector for each chain are also indicated

two chains with each other. Although there are three 1D chains (see Sect. 3.1.1.2), only two 2D structures can be constructed out of them with two molecules in the unit cell [11]: a hexagonal arrangement, Fig. 3.3, and a close-packed structure, Fig. 3.4.

Both arrangements are planar. The hexagonal monolayer is more stable in energy and has a higher symmetry in comparison with the close-packed structure. Both structures have six hydrogen atoms per unit cell involved in the formation of hydrogen bonds, but in the close-packed case, see Fig. 3.4, there are structural features that weaken its H-bonding. Exactly like in the case of close-packed Chain3, which this

Table 3.3 Energies and geometrical parameters of the melamine chains

	Chain1	Chain2	Chain3
E_{stab} (eV)	−0.87	−0.85	−0.70
E_{int} (eV)	−1.13	−1.10	−0.99
E_{def} (eV)	0.25	0.25	0.29
N–H...N distance within the unit cell (Å)	2.90	2.92	3.01
N–H...N distance between the unit cells (Å)	2.89	2.91	3.06
$N-\widehat{H}\ldots N$ angle within the unit cell (°)	177.7	178.3	177.6
$N-\widehat{H}\ldots N$ angle between the unit cells (°)	179.6	178.3	178.0
Lattice vector A_1 (Å)	10.70	12.26	7.76

The stabilisation and interaction energies include the BSSE correction. The last five rows show the N–H...N distances and the N–H...N bond angles within and between unit cells, as well as the lattice vectors of the one-dimensional chains

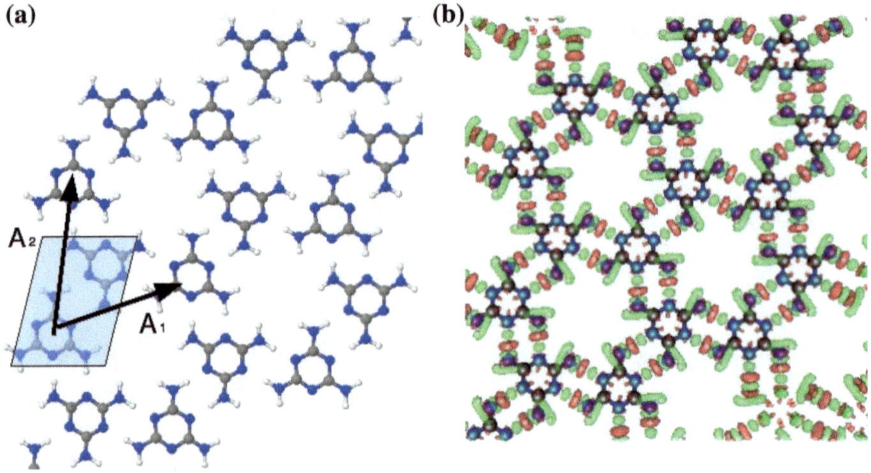

Fig. 3.3 The geometry **a** of the hexagonal melamine monolayer and its electron density difference plot **b** corresponding to ±0.01 electrons/Å^3

network is based on , one amino group (out of three) of each molecule participates in two hydrogen bonds simultaneously so that the corresponding nitrogen donor is unable to provide enough electron density to the bonds. Also, this structure is very dense and some hydrogen atoms are positioned very close to each other thus forcing them to move out of the molecular plane, see Fig. 3.4c, to relieve the strain in the system. Due to this reason, the close-packed structure is less stable. This conclusion is also supported by comparing the stabilisation energies of the 2D structures with the predicted value of -1.32 eV from the (planar) dimer energies, considering that there are six hydrogen bonds per unit cell. The predicted value agrees well with the stabilisation energy of the hexagonal structure, Table 3.4, but it is much bigger than the close packed stabilisation energy.

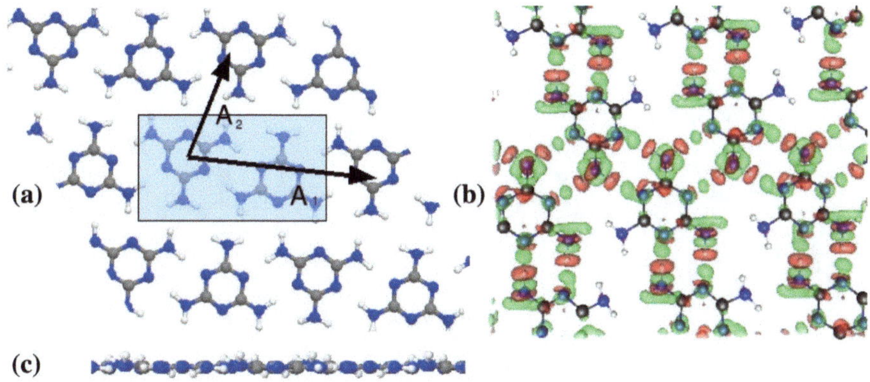

Fig. 3.4 The *top* (**a**) and *side* (**c**) views of the close-packed melamine monolayer and its electron density difference plot (**b**) corresponding to ±0.01 electrons/Å^3

Table 3.4 Stabilisation, interaction and deformation energies and geometrical parameters of the melamine 2D networks with two molecules in the unit cell

	Hexagonal	Close-packed
E_{stab} (eV)	−1.29	−0.99
E_{int} (eV)	−1.67	−1.29
E_{def} (eV)	0.38	0.30
N–H...N distance within the unit cell (Å)	2.88	3.05
N–H...N distance between the unit cells (Å)	2.88	2.98
$N-\widehat{H}\ldots N$ angle within the unit cell (°)	178.8	170.4
$N-\widehat{H}\ldots N$ angle between the unit cells (°)	178.8	173.1
Lattice vector A_1 (Å)	10.66	12.00
Lattice vector A_2 (Å)	10.66	7.68
$\widehat{A_1A_2}$ angle between A_1 and A_2 (°)	60	65.3

The stabilisation and interaction energies include the BSSE correction. The N–H...N distances and the N–H...N angles within the unit cell and between two different unit cells, the lattice vectors of the networks and the angle between them are also shown

Both hexagonal [3, 5, 11] and close packed [11] structures have been experimentally observed and the geometrical properties of these two structures are in good agreement with the experimentally measured ones [11].

In this section we have described the systematic approach used to build the 2D structures based on the melamine dimer. Note that these are gas-phase calculations, which do not include the effect of the substrate. The approach presented here is general and applicable to all our systems. In the following sections our method will be extended towards building structures based on more than two molecules in the unit cell.

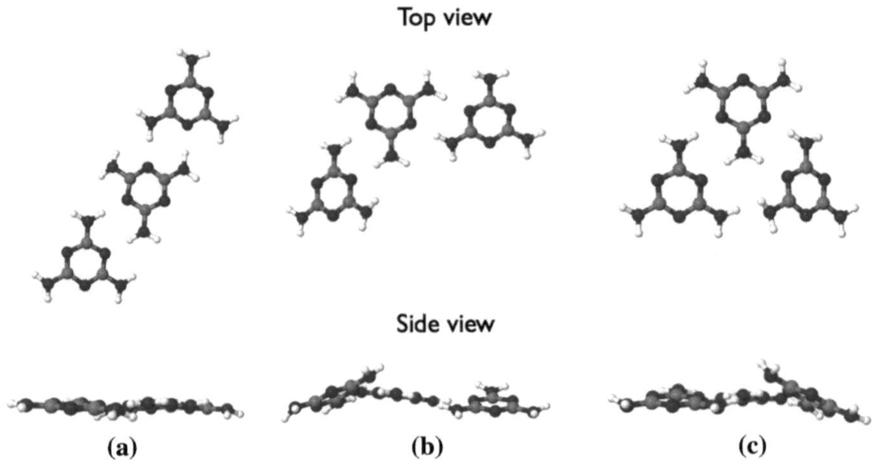

Fig. 3.5 The *top and side views* of the three possible trimers

3.1.2 Trimers

In this section we consider melamine trimers and all possible structures which one can construct based on three molecules in the unit cell. Such cells are built starting from the dimer. Firstly, the dimer binding sites available to form hydrogen bonds with the third molecule are identified as in the previous section. The third molecule in each trimer would have the same orientation as the first one, i.e. it is rotated by 60° with respect to the second melamine (to which it is immediately connected by a double hydrogen bond). Taking into account the symmetry of the dimer, there are four distinct sites available on a dimer to build the trimer cell. However, we find that only three trimers are possible, as shown in Fig. 3.5.

The trimers are stabilised by four hydrogen bonds, so that the predicted stabilisation energy for either of the trimers is −0.88 eV, based on the planar dimer structure, and −0.96 eV if it is based on the non-planar one. Although the starting geometries were in all cases planar, the relaxed structures were all non-planar. This fact is reflected in the N–H...N angles in the trimers being different from 180°, as shown in Table 3.5.

All trimers have similar stabilities. Trimer2 has a similar structure to Chain1 based on the melamine dimer; Trimer1 is similar to Chain3, while Trimer3 has the structure similar to Chain3. The stabilisation energies of Trimer3 and Chain3 (per molecule) are also similar (−0.32 and −0.35 eV, respectively, see Tables 3.3, 3.5).

Following the same logic as we used for the chains based on the dimer unit cell, all available binding sites on the periphery of the trimers can be identified, and 1D chains based on all trimers can then be built. Taking into account the symmetry of the trimers, eight chains can be constructed, each having six hydrogen bonds per unit cell as shown in Fig. 3.6.

Table 3.5 Stabilisation, interaction and deformation energies and geometrical parameters of the melamine trimers

	Trimer1	Trimer2	Trimer3
E_{stab} (eV)	−0.98	−0.99	−0.96
E_{int} (eV)	−1.19	−1.21	−1.15
E_{def} (eV)	0.21	0.22	0.19
$d_{N-H...N}$ (Å)	2.89 ± 0.01	2.90	2.94 ± 0.04
$\widehat{N-H...N}$ (°)	177.4 ± 0.2	177.8 ± 0.5	178.1 ± 0.8

The stabilisation and interaction energies include the BSSE correction. The last rows show the average N–H...N distance and the N–H...N angle in the three structures

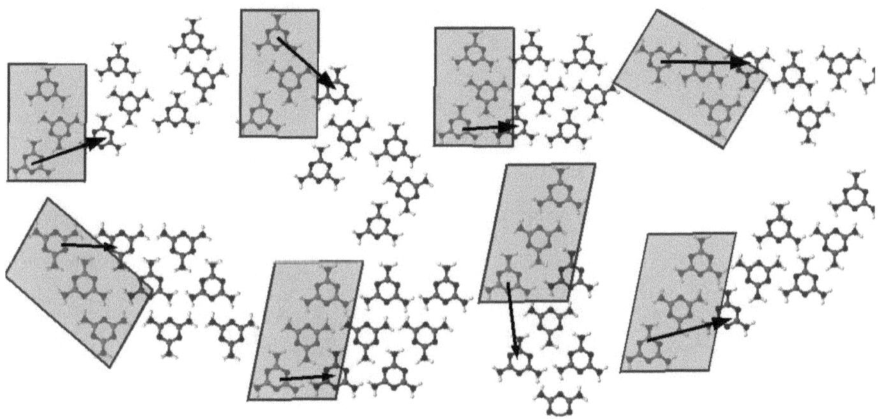

Fig. 3.6 Models of eight 1D chains based on the trimer unit cells

Stabilisation energies (per cell) of these chains can be predicted as before from the dimer energies and they are −1.32 eV for the planar and −1.44 eV for the non-planar trimers, respectively. However, our analysis shows that it is not possible to connect the trimer chains together to form 2D structures, because all chains have edge molecules with the same orientation. To construct 2D structures, a fourth molecule is required which is 60° rotated.

Thus, to have 1D chains and 2D networks based on the same unit cell, a half of the molecules in the cell should be rotated by 60° with respect to the others. This means that in order to be able to create 2D structures stabilised by hydrogen bonds, we need to use unit cells with *even* numbers of molecules, i.e. it is not possible to do it with trimers. The trimer structures, however, are necessary as a starting point to build tetramers which will be used in the next section as the building blocks for constructing chains and monolayers with four molecules in the unit cell.

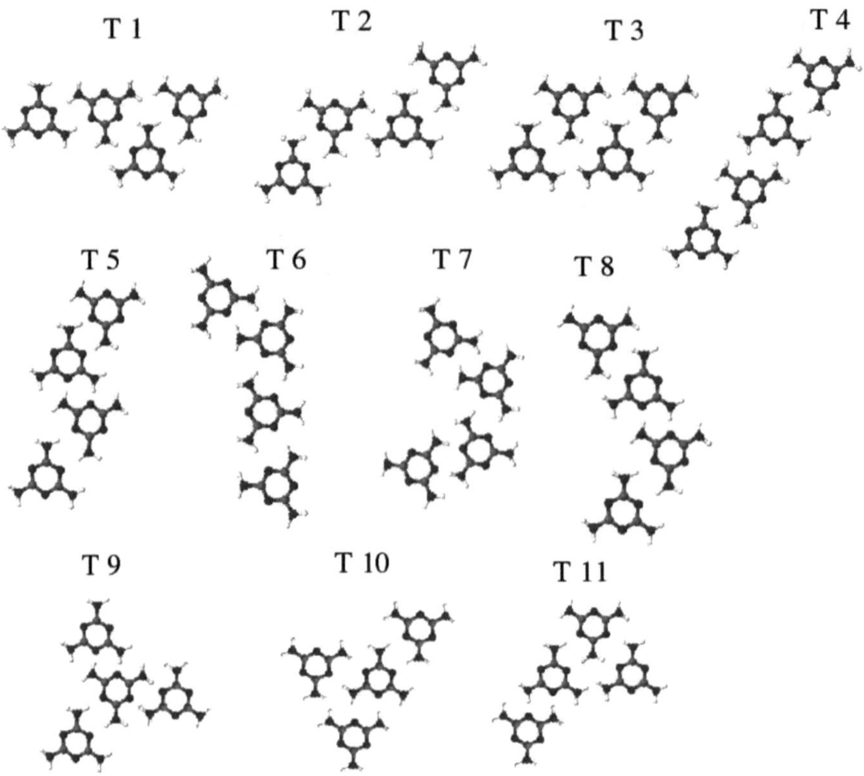

Fig. 3.7 The geometries of all possible tetramer structures

3.1.3 Tetramers

In this section we will present the results on melamine tetramers, and the corresponding chains and networks based on four molecule unit cells. In the same way as described above, we identified the binding sites in the trimers, available for connecting with another melamine molecule, and found that 11 tetramers can be built as shown in Fig. 3.7. All structures, once relaxed, become non-planar. To construct periodic structures, we have to consider all tetramers in their ability to connect to each other in 1D and 2D periodic geometries. Therefore, a very large number of structures may be possible. In practice, however, a much smaller number of structures is derived. In particular, it is not possible to set up a super-structure with tetramers T9 or T10 as a unit cell as they have three molecules with the same orientation at their edges, and thus no sites are available to form hydrogen bonds with the same cluster. On the other hand, other tetramers such as T2, T3, T4 and T5 can be described as two identical dimers connected to each other and their combination may lead to 1D and 2D structures in which the periodicity is actually based on unit cells with two molecules, and we have already considered such structures in Sect. 3.1.1. Finally,

Table 3.6 Stabilisation, interaction and deformation energies and geometrical parameters of the first four melamine tetramers

	T1	T2	T3	T4
E_{stab} (eV)	−1.51	−1.38	−1.48	−1.46
E_{mol} (eV)	−0.38	−0.35	−0.37	−0.37
E_{int} (eV)	−1.84	−1.76	−1.81	−1.78
E_{def} (eV)	0.33	0.38	0.33	0.32
N–H...N distance (Å)	2.89 ± 0.03	2.89 ± 0.01	2.92 ± 0.02	2.90 ± 0.02
$\mathrm{N-\widehat{H}\ldots N}$ angle(°)	177 ± 1	177 ± 1	177 ± 2	178 ± 1

The stabilisation energy per molecule, E_{mol}, is also given. This, as well as stabilisation and interaction energies include the BSSE correction. The last rows show the mean N–H...N distance and the N–H...N angle

starting from different tetramers, identical 2D structures can be constructed because of the very high symmetry of the melamine molecule.

Table 3.6 describes the energies and geometries of the first four melamine tetramers. Only these four tetramers are considered here because they are sufficient to build all the possible 1D and 2D structures, as we shall show in the next section.

Each tetramer has six hydrogen bonds, so that the predicted stabilisation energies for the tetramers are −1.32 and −1.44 eV for the planar and non-planar configurations, respectively. All structures have similar stabilisation energies, shown in Table 3.6, which are slightly higher than the predicted energies. This indicates a cooperative effect, i.e. neighbouring hydrogen bonds affect favourably each other, so that the redistribution of the electron density, which appears as the alternating "kebab"-like pattern on one bond, electrostatically favours a "kebab"-like structure on the other, the so-called resonant assistant hydrogen bonding (RAHB) [12]. The RAHB effect is known to play a role in some systems stabilized by hydrogen bonding [10, 13–15]. To characterize this collective effect, it is convenient to perform also an approximate calculation in which the binding energy of the whole complex system is estimated as a simple sum of binding energies of each pair of molecules involved. If for instance the DFT calculated binding energy of the whole system is lower (more negative) than the sum of dimer binding energies, then this would indicate on the existence of the positive RAHB effect, i.e., that the hydrogen bonds effect favorably each other in the system.

By connecting the tetramers together along one directions, all possible chains can be constructed. We find that there are 51 different 1D structures possible. We do not show them here to save space and we will instead concentrate, in the next section, on all possible 2D monolayers which are the main concern of this study.

3.1.3.1 Two-Dimensional Structures Based on the Melamine Tetramers

By putting tetramer based chains parallel to each other, all possible 2D periodic monolayers can be constructed. Because of the high symmetry of the melamine molecule,

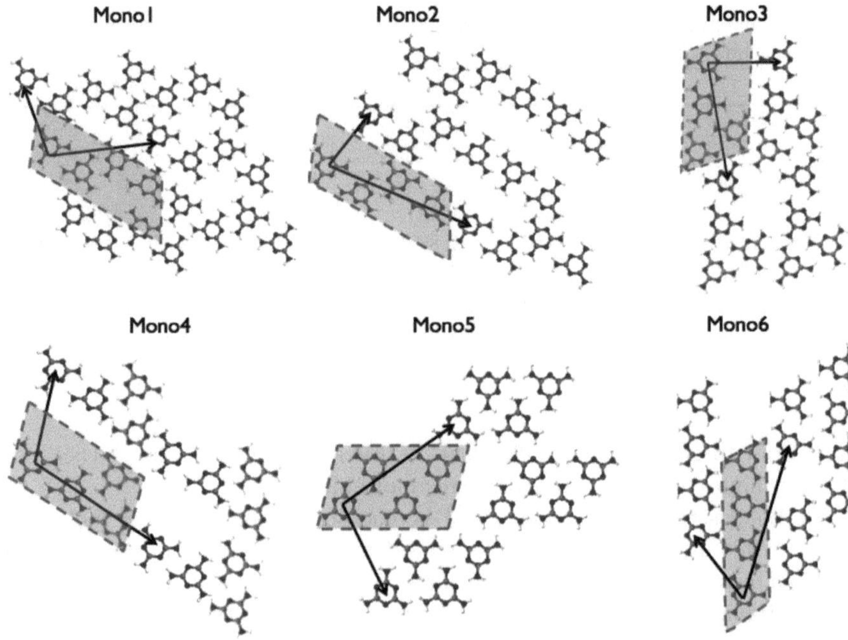

Fig. 3.8 Six monolayers based on four molecule per unit cells. Unit cells and the corresponding lattice vectors are also indicated

the large number of chains results in only 8 possible distinct 2D monolayers. These include the two already considered in Sect. 3.1.1.3, based on the two molecule unit cell (the dimer), which may be obtained from tetramers T2–T5 consisting of two identical dimers. In fact, in order to build up all 8 monolayers, it is sufficient to use only the first four tetramers T1–T4.

The six new 2D structures based on the melamine tetramers are shown in Fig. 3.8, and their energetic and geometrical characteristics are given in Table 3.7.

We find that when using four-molecule cells, it is possible to obtain 2D structures with more complicated shapes and with larger pores than even in the hexagonal network shown in Fig. 3.3. In the case of the two-molecule cell structures, each molecule is three-fold coordinated and has all the sites employed in the hydrogen bond formation. In the networks Mono2–Mono6 based on the tetramer unit cells, some of the molecules are only two-fold coordinated and thus have some of their binding sites unused. These monolayers all have similar stabilisation energies, see Table 3.7, which can be compared with the predicted binding energy of −2.22 eV calculated for 10 hydrogen bonds (five dimer connections) per unit cell. We see that in all cases the actual energies are slightly higher the predicted value indicating the RAHB effect. Only one of the tetramer-based structures, namely Mono1 shown in Fig. 3.8, has every molecule surrounded by three other molecules, and, sure enough, this is the most stable tetramer-based monolayer, see Table 3.7. Its stability is also

Table 3.7 Stabilisation, interaction and deformation energies (per cell) and geometrical parameters of the melamine 2D networks based on tetramers

	Mono1	Mono2	Mono3	Mono4	Mono5	Mono6
E_{stab} (eV)	−2.72	−2.31	−2.35	−2.34	−2.29	−2.33
E_{mol} (eV)	−0.68	−0.58	−0.59	−0.58	−0.56	−0.58
E_{int} (eV)	−3.31	−2.87	−2.91	−2.83	−2.81	−2.88
E_{def} (eV)	0.59	0.56	0.56	0.49	0.52	0.54
d_{N}−H...N (Å)	2.92 ± 0.04	2.90 ± 0.04	2.90 ± 0.04	2.90 ± 0.06	2.92 ± 0.06	2.92 ± 0.05
N−$\widehat{H}$...N (°)	175 ± 4	176 ± 2	176 ± 2	176 ± 2	176 ± 2	177 ± 1

The stabilisation and interaction energies include the BSSE correction. The stabilisation energy per molecule, E_{mol}, is also given. The last rows show the mean N–H...N distance and the N–H...N angle in the monolayers

higher than the predicted value of −2.64 eV based on 12 hydrogen bonds per cell, so that the RAHB effect is at work here as well. In fact, Mono1 is the highest stability network we have found; indeed, it is more stable even than the hexagonal structure considered in Sect. 3.1.1.3, as the binding energy of Mono1 per two molecules is −1.36 eV, compare with Table 3.4.

3.1.4 Comparison with Experimental Data

Assuming that there are only two molecules in the two-dimensional unit cell, we have found, using our systematic method, (Sect. 3.1) that *only two* monolayers are actually possible: (i) the hexagonal structure shown in Fig. 3.9c and (ii) a close-packed arrangement, whose relaxed structure is shown in Fig. 3.9d. The small number of possibilities is explained by the high symmetry of the melamine molecule. If the hexagonal structure is *non-chiral*, the close-packed one is *chiral*, as the latter monolayer, when flipped over, cannot be made identical to the original one by means of any rotations and/or translations.

The calculated lattice parameters of the two structures, which are shown in Table 3.8, compare well with those measured experimentally. We have also checked that the relaxed structures match well with the observed ones if overlayed on their images, suggesting that they represent the correct models for the observed monolayers.

By considering a larger number of the molecules in the unit cell, more monolayers can be constructed, as, is done in Sects. 3.1.2 and 3.1.3. Now, assuming four molecules in the cell, we find ten possible unit cells which can be used to construct eight different monolayers. One example is the hexagonal structure corresponding to four molecules in the unit cell and it is shown in Fig. 3.10b. This structure is quite peculiar: it consists of elements of both monolayers (which were based on two-molecules cells) considered above, see Figs. 3.3 and 3.4. Indeed, one can see hexagons and close-packed rows of molecules with some of their amino groups participating in

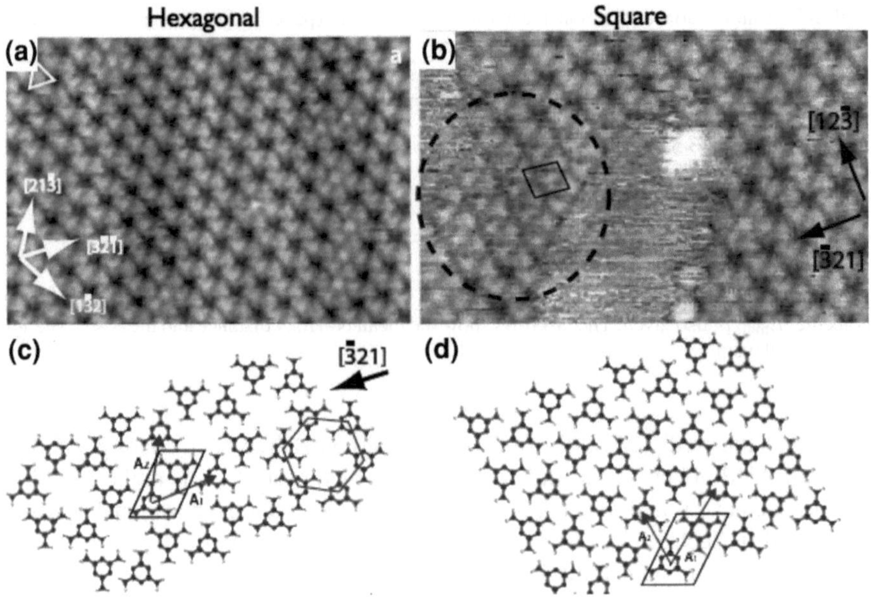

Fig. 3.9 STM image of **a** the hexagonal and **b** close-packed melamine domain on a Au(111)-$(22 \times \sqrt{3})$ surface (**a** $13 \times 9\,\mathrm{nm}^2$; $V_s = -1.1\,\mathrm{V}$, $I_t = 0.5\,\mathrm{nA}$ and **b** $12 \times 8\,\mathrm{nm}^2$, $V_s = -1.1\,\mathrm{V}$, $I_t = 0.5\,\mathrm{nA}$.); **c** the corresponding molecular model consisting of hexagons and **d** square arrangement

Table 3.8 Comparison between the theoretical (assuming two-molecule unit cells) and the experimental lattice parameters for the two monolayers

	Lattice vector A_1 (Å)	Lattice vector A_2 (Å)	Angle
Hexagonal (theory)	10.66	10.66	60°
Hexagonal (experiment)	10 ± 1	10 ± 1	$60° \pm 2°$
Close-packed (theory)	12.00	7.68	65.3°
Close-packed (experiment)	10 ± 1	8 ± 1	$63° \pm 4°$

two double hydrogen bonds at the same time. In addition, in this structure, as well as in the close-packed monolayer of Fig. 3.9d, some of the hydrogens are positioned very close to each other forcing them to move out of the molecular plane to relieve the strain in the system. One can also appreciate that this structure is quite stable with the stabilisation energy of $-1.21\,\mathrm{eV}$ per two molecules (this is only slightly less stable than the hexagonal structure of Fig. 3.3). This four-molecule structure can serve as an atomistic model for the close-packed boundary region shown in Fig. 3.4.

In the next section the same kind of approach will be applied to more complicated molecules such as PTCDA, PTCDI, NTCDA, NTCDI and to mixed phases of melamine and PTCDA/PTCDI.

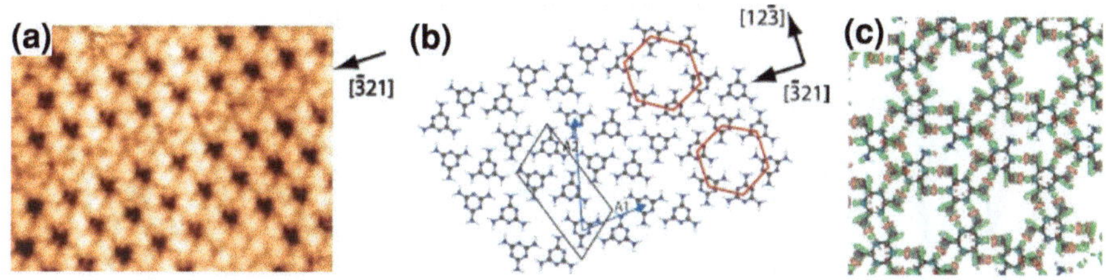

Fig. 3.10 **a** STM-image of a melamine domain on a Au(111) surface ($V_s = -1.1$ V, $I_t = 0.5$ nA, $8 \times 6\,\mathrm{nm}^2$); **b** molecular model and **c** the electron density difference plot of the region along the arrow in **a**. (The theoretical lattice vectors are: $A_1 = 10.65$ Å, $A_2 = 17.69$ Å; angle between them 51.3°)

3.2 PTCDA

In this section we shall consider our theoretical method in application to all possible gas-phase periodic structures formed by PTCDA molecules. All structures based on two molecules per unit cell will be described in detail first. Then we shall discuss what one can expect when going beyond this limitation. We start by describing in detail the approach used and shall present the PTCDA dimer; this will then be followed by considering the construction of all superstructures.

3.2.1 PTCDA Dimers

The starting point in our analysis is the identification of the binding sites in the PTCDA molecule, see Fig. 3.11. Six possible non-equivalent binding sites have been identified. Sites 2, 4 and 5 are exclusively of donor type; they are composed of carbon-hydrogen groups and are able to form either double or triple hydrogen bonds. Two of the binding sites are exclusively acceptors (3 and 6) and are composed by oxygens enabling them to form three or two hydrogen bonds as well. There is also a third kind of binding site (site 1) with both donor and acceptor atoms (it is composed of an oxygen and hydrogen atoms able to form a double hydrogen bond); there are four such bonds along the molecule perimeter.

Although the PTCDA molecule has many binding sites, because of the symmetry of the molecule, only a limited number of distinguishable dimers are possible. In fact, in combining two PTCDA molecules, four dimers are obtained and are shown in Fig. 3.12. Three of these dimers (D1, D2 and D4) are very similar, as they are constructed by placing one PTCDA molecule perpendicularly with respect to the other and shifting along its longer side. Dimer D1 is formed by aligning two hydrogen atoms in the centre of the long side of one molecule with the three oxygens of the other. Dimers D2 and D4 are obtained by shifting the molecules with respect to each other as compared with the D1 configuration. The final dimer structure D3 is

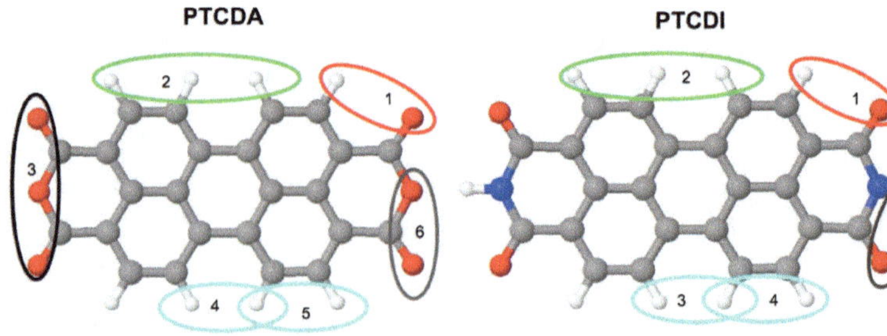

Fig. 3.11 PTCDA and PTCDI molecules and their non-equivalent binding sites shown by *numbered ovals*

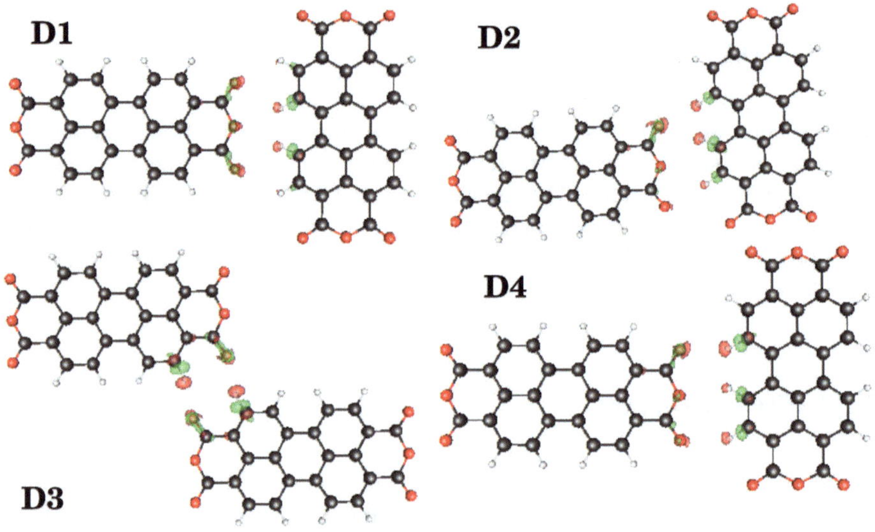

Fig. 3.12 The geometries of the four possible PTCDA dimers shown together with the electron density difference plots corresponding to ± 0.01 electrons/Å^3. The *green* surfaces correspond to the regions of positive electron difference (excess) and the *red* areas correspond to the regions of negative electrons density difference (depletion)

obtained when the two PTCDA molecules are connected in a "head-and-tail" like structure, see Fig. 3.12, using the binding site 1.

The four dimers are found to relax into stable geometries. As expected, the stabilisation energies of the dimers shown in Table 3.9, are found to be rather small and similar; this also means that it could be energetically easy to switch between the D1, D2 and D4 dimers. In the cases of similar D1, D2 and D4 dimers all three oxygens of one molecule participate, however, a different number of hydrogen atoms (between 2 and 3) participate in the other.

As evidenced by the electron density difference plots in Fig. 3.12 and the binding energies, in all cases the dimers are very weak. By a small margin, the most stable

Table 3.9 Stabilisation, interaction, deformation and BSSE energies (in eV) of the four PTCDA dimers

	D1	D2	D3	D4
E_{stab}	−0.25	−0.23	−0.26	−0.27
E_{def}	0.01	0.01	0.02	0.02
E_{int}	−0.26	−0.24	−0.28	−0.29
E_{BSSE}	0.10	0.08	0.13	0.08

The stabilisation and interaction energies include the BSSE correction in each case

dimer is D4, remaining still very weak. This is peculiar since the most developed "kebab" structure is displayed by the dimer D3. Dimer D4 in the end appears to be marginally stronger due to the overall effect of having three (weak) hydrogen bonds.

One would expect stronger hydrogen-bonds when the oxygen atom is involved due to its high electronegativity, however, from the chemical point of view, in all cases the hydrogen donors are carbon atoms that are not enough electronegative to contribute sufficiently into the hydrogen bond formation. Small deformation energies E_{def}, shown in Table 3.9, underline the rigidity of the PTCDA molecule, including its C-H groups.

Once all the possible dimers are constructed, the next step is to identify the binding sites available to connect the PTCDA dimers into a one-dimensional (1D) chain. In the next section the possible chains based on the PTCDA pairs are presented.

3.2.2 One-Dimensional Chains Based on the PTCDA Pairs

In this section we present the 1D structures that can be built using a PTCDA dimer as an unit cell. As before, we start from the identification of the binding sites available on the periphery of the dimer and for the connection with more dimers, see Fig. 3.12. We build 1D chains by connecting dimers in a chain-like manner along one direction. The chains are obtained by adding a second dimer to a certain binding site of the first dimer, then a third dimer is added to the same site of the second one, and so on. In this way, by using free dimer binding sites 30 different chains can be constructed. However, as many chains are geometrically similar (mainly due to similarity of the dimers D1, D2 and D4), one can group them into families of the geometrically similar chains and then calculate the stabilisation energy for a single representative of each of them. The PTCDA chains therefore were classified into 10 distinct families, and one chain from each family was selected and its energy calculated with DFT; their relaxed structures are presented here in Fig. 3.13 and the energies are given in Table 3.10.

In the previous Sect. 3.2.1 we found that the PTCDA dimers are very close in energy. Therefore, it does not come as a surprise that the calculated stabilisation energies of chains constructed from them are also very close, spanning a narrow energy interval between −0.47 and −0.61 eV, i.e. the largest energy difference of only 0.14 eV is found here for the different chains, see Table 3.10.

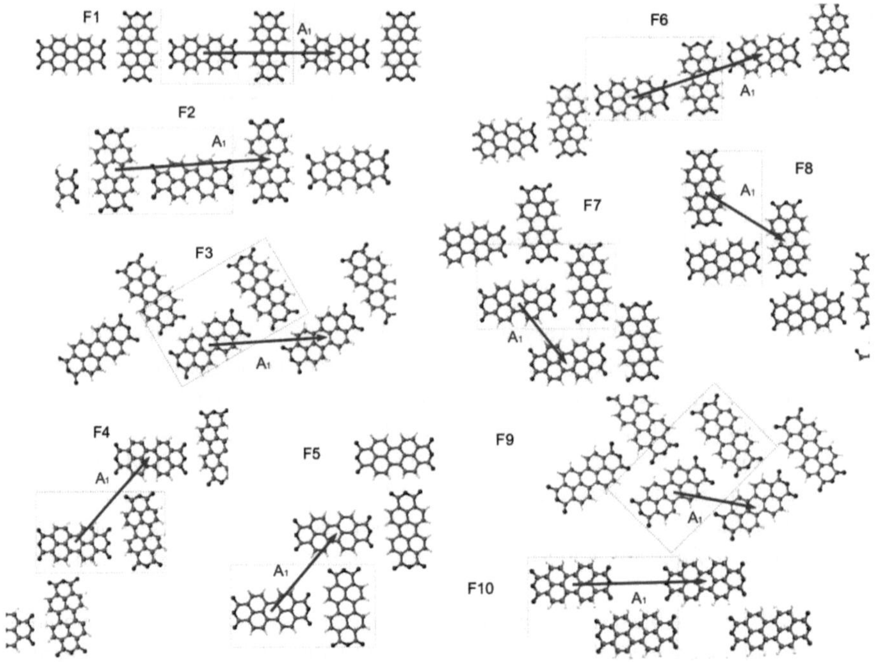

Fig. 3.13 Relaxed geometries of the 10 PTCDA chains, named from F1 to F10 and representing 10 distinct families of chains. The lattice vector $\mathbf{A}_1$ is shown in all cases and the unit cell is indicated by a *dashed box* for convenience

Table 3.10 The calculated (using SIESTA) and evaluated (as a sum of dimer energies) stabilisation energies (in eV) of all ten PTCDA chains

	F1	F2	F3	F4	F5
E_{stab} (calc.)	−0.47	−0.50	−0.56	−0.52	−0.60
E_{stab} (eval.)	−0.48	−0.50	−0.44	−0.44	−0.46
	F6	F7	F8	F9	F10
E_{stab} (calc.)	−0.56	−0.56	−0.48	−0.61	−0.48
E_{stab} (eval.)	−0.44	−0.50	−0.44	−0.50	−0.52

We have also presented in the table the evaluated stabilisation energies, as described in Sect. 2.5, which are sums (per unit cell) of the corresponding dimer energies. One can notice that the evaluated and the calculated (with DFT) stabilisation energies are generally very close; small differences of up to −0.14 eV are found for some of the chains where there is an additional cooperative effect due to multiple hydrogen bonds, e.g. chain F9, see Fig. 3.13; chains F5 and F9 are the most stable ones. These chains are very similar in geometry, and have the maximum number of hydrogen and oxygen atoms involved in the chain formation.

Table 3.11 The evaluated (by summing up dimer energies) and calculated (with DFT) stabilisation energies of eight representative PTCDA monolayers

	Square	Herringbone						Brick wall
Index	MON1	MON2	MON3	MON4	MON5	MON6	MON7	MON8
E_{stab} (eval.)	−1.00	−1.00	−1.10	−1.10	−0.72	−0.78	−0.70	−1.04
E_{stab} (calc.)	−1.19	−1.17	−1.19	−1.34	−0.7	−1.01	−0.61	−0.89
E_{BSSE}	0.42	0.40	0.42	0.58	0.28	0.47	0.26	0.50

BSSE corrections are also shown for each structure. For the ease of comparison, energies for the MON8 structure containing a single molecule in the unit cell, were doubled

The next step is to combine the chains and obtain all possible 2D structures.

3.2.3 Two-Dimensional Structures Based on the PTCDA Pairs

In order to build all possible 2D structures, we considered each chain separately and identified again the binding sites on either side of the chain. Out of 30 1D chains available (see Sect. 3.2.2), it is possible to generate 42 2D structures in which molecules are attached to each other via the PTCDA dimers connections.

Such a rather significant number of possible monolayers is due to a large number of dimers PTCDA molecules can form (however, because of a very similar nature of the dimers D1, D2 and D4, many monolayers look alike and their energetics and the structural motif are expected to be similar as well). Therefore, the same method as in the previous section has been used to simplify our discussion. Since we know all the dimers binding energies, the energy (per unit cell) of each monolayer can be evaluated by summing up all the relevant dimer energies. Then it is possible to classify the monolayers into families of "similar" monolayers taking into account both the geometry and the evaluated energy of each monolayer. In this way, eight families of monolayers have been identified and hence eight prototype monolayers (one from each family) have been considered for further DFT study. These representative structures, relaxed with our DFT method, together with the corresponding lattice vectors, are shown in Fig. 3.14. Note that the structure MON8 is based on a single molecule in the unit cell; however, to simplify the forthcoming discussion, for this structure we consider a supercell with two molecules as indicated in Fig. 3.14.

We have also considered separately another monolayer MON9 shown in Fig. 3.15. Although this structure formally belongs to the second family of structures, it was found to be convenient to separate it out since it was found to be the best match for the domino phase observed in the STM image in Fig. 3.16b.

The stabilisation energies for the selected eight monolayers are given in Table 3.11, while the density difference plots for three of them are shown in Fig. 3.17. The lattice vectors in each case are also provided in Table 3.12.

In the present case, the evaluated stabilisation energies for monolayers MON1-MON4 and MON6 based on PTCDA pairs are smaller than the actual calculated

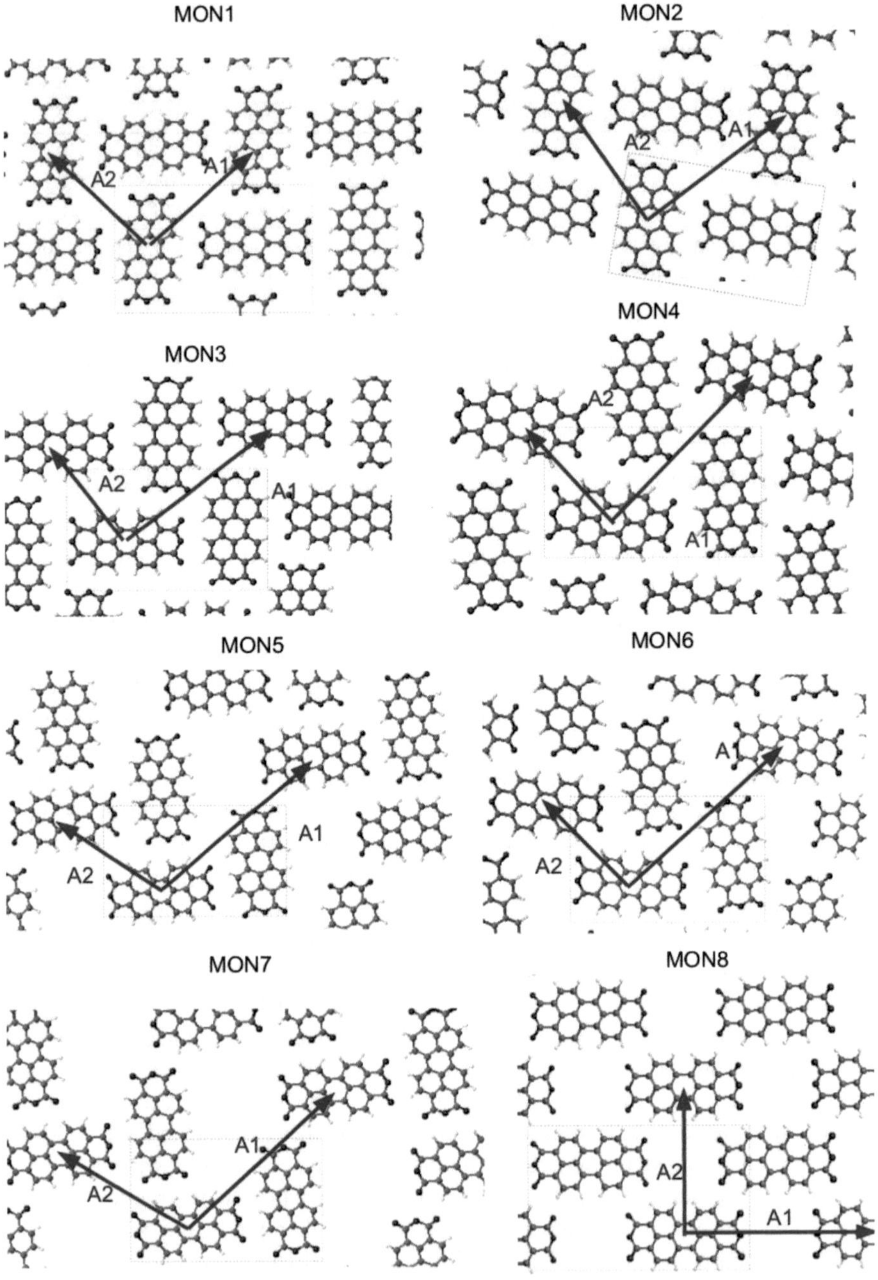

Fig. 3.14 Eight PTCDA monolayers, representing eight distinct families of them, relaxed with our DFT method. Both the lattice vectors, $\mathbf{A}_1$ and $\mathbf{A}_2$, and the unit cell in each case are indicated

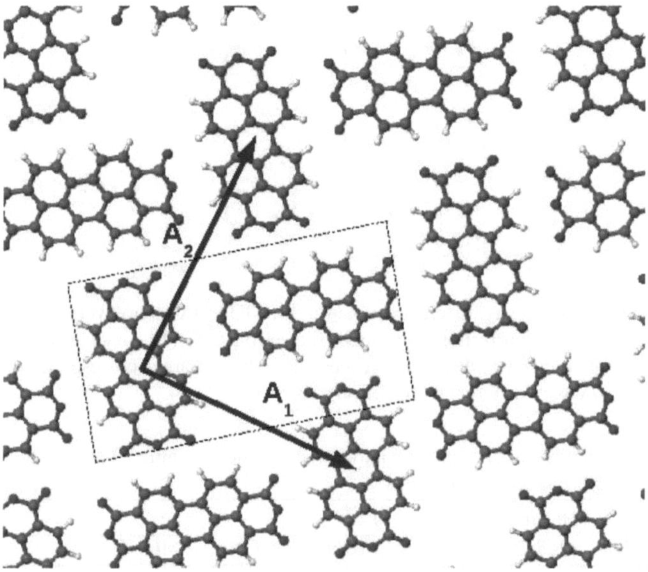

Fig. 3.15 The monolayer MON9. Both the lattice vectors, $\mathbf{A}_1$ and $\mathbf{A}_2$, and the unit cell are indicated

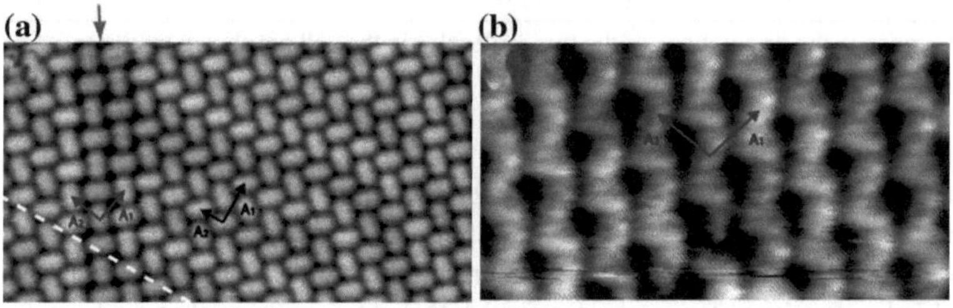

Fig. 3.16 STM images of the observed new PTCDA networks on the Au(111) surface. **a** The narrow stripe of the square phase providing a buffer for two shifted herringbone phases on the left and right of it. The *big black arrow* at the *top* of the image indicates the square phase. The two lattice vectors of the two square and herringbone phases are explicitly indicated by *blue and black arrows*, respectively. The scanned area is $210 \times 150\,\text{Å}^2$. The tunneling current $I_t = 0.3$ nA and the applied voltage $V_s = 1.5$ V; **b** the domino phase; the two lattice vectors are indicated by the *blue arrows*. The scanned area is $105 \times 70\,\text{Å}^2$. The tunneling current $I_t = 0.7$ nA and the applied voltage $V_s = -0.3$ V [16] (Colour figure online)

(with DFT) energies. This is because the geometries changed after relaxation with the DFT method and cooperative effects must thus be playing a role. The MON4 is the most stable monolayer and the extra stabilisation energy can be identified as a result of the formation of additional hydrogen bonds in the structure. This is most clearly seen in the density difference plot in Fig. 3.17. The least stable monolayer is MON7 with the corresponding density difference plot also shown in Fig. 3.17 for

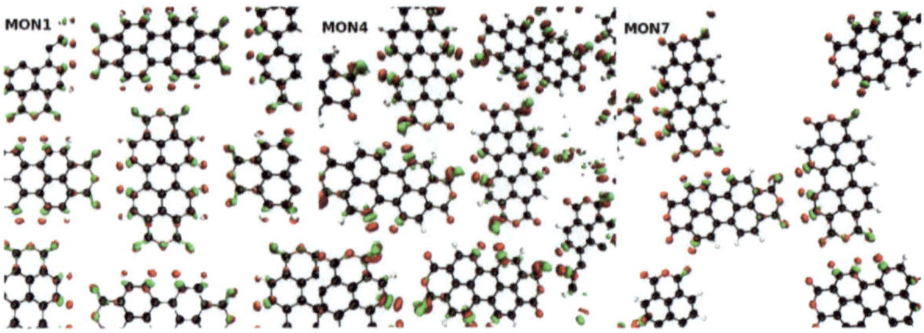

Fig. 3.17 The electron density difference plots for three PTCDA monolayers, MON1, MON4 and MON7, with the contours corresponding to ±0.01 electrons/Å^3. The *green* surface corresponds to the regions of positive electron difference (excess) and the *red* areas correspond to the regions of negative electrons density difference (depletion)

Table 3.12 The lengths (in Å) of the two lattice vectors and the angle γ between them (in degrees) for the selected eight PTCDA monolayers

	Square	Herringbone						Brick wall
Index	MON1	MON2	MON3	MON4	MON5	MON6	MON7	MON8
A_1	16.47	17.3	20.1	19.6	23.13	22.28	22.32	19.54
A_2	16.26	15.5	12.74	12.71	13.78	12.41	16.72	15.6
γ	90.1	89.6	87.6	90.1	102.1	96.7	108.0	90
τ	90	87.6	89.8	79.6	88.6	79.6	87.9	0

The difference in the orientations of the two molecules in the cell are shown by the angle τ (also in degrees)

comparison; in every molecule has only three "contacts" with other molecules (less than the other structures) and the corresponding hydrogen bonds are very weak. As an example of a monolayer with intermediate stabilisation energy, we show in Fig. 3.17 the density difference plot for MON1. One can see that each molecule is "connected" to four neighbours, however, all hydrogen bonds are rather weak.

The monolayer MON1 corresponds to the square phase observed experimentally [17–21], MON2-MON7 monolayers present the herringbone phase [17, 19–28], while MON8 is the brick wall phase [24, 29]. Lattice vectors of some of the observed monolayers on different substrates (defined in the same way as in Fig. 3.14) are shown in Table 3.13 for comparison. The geometrical characteristics of the square and herringbone phases observed in this work (see Sect. 1.1.2) are also given. We see that some of our calculated gas-phase monolayers compare reasonably well with the experimental geometries, e.g. the square monolayer MON8 is very similar to the one observed on the Ag-terminated Si(111) surface [17]. We also observe, in Fig. 3.14 and Table 3.12, considerable variations in the calculated possible herringbone phases which comprise 6 families of structures differences in their lattice vectors and the angle τ between the molecules in the unit cell. This variation is possible, in the first place, due to the similarity of the D1, D2 and D4 dimer connections between PTCDA molecules in the perpendicular orientation. Monolayers MON3 and MON4 are

Table 3.13 Geometrical characteristics of some of the experimentally observed PTCDA monolayers

	Substrate	References	A_1	A_2	γ	τ
Square	Si/Ag(111)	[17]	16.4	16.4	90.6	90
	Au(111)	[16]	17.0	17.0	88	88
Herringbone	Si/Ag(111)	[17]	20.01	11.55	90	88.4
	Cu(111)	[22]	21.6	13.6	90	90
	Au(111)	[23]	20.0	12.0	90	86
	Au(111)	[16]	19.4	12.5	88	85
	Ag(111)	[24]	19.0	12.6	89	80
	KBr(001)	[25]	19.91	11.96	90	83.2
Brick wall	Ag(110)	[24]	17.31	16.33	90	0
Domino	Au(111)	[16]	15.5	15.3	105	90

See Table 3.12 for the explanation of the notations. In some cases the corresponding data were missing in the quoted papers and hence was obtained directly from their images

similar to the observed ones in [17, 23–25] and in this work, while monolayers MON5 and MON6 may be compared with the one observed in [22]. Several herringbone phases have also been observed in some studies on the same sample [20–22, 25]. The low density monolayer MON7 has very large lattice vectors and it has never been observed, most likely, due to its relatively low stability. The brick wall monolayer MON8 in the gas phase, as calculated by us, has noticeably different lattice vectors as compared to that observed on the (corrugated) Ag(110) surface [24].

3.2.4 Going Beyond Two Molecules Per Cell

We note that the hexagonal structure seen in [17] has not appeared in our analysis above. This is because, so far we have limited our study to the structures based on two molecules in the unit cell only. However, it can easily be seen that this particular hexagonal structure, containing 4 molecules in the unit cell, can be obtained by placing two different chains parallel to each other, namely F1 and F6 from Fig. 3.13, and then repeating the arrangement periodically in a sequence ...-F1-F6-F1-F6-.... Therefore, this hexagonal structure would necessarily be considered (alongside other possible arrangements) if our method was to be applied with the assumption of 4 molecules in the cell.

One question still remains, however: it is not a priori obvious that such an arrangement would be stable. In order to check this point, we build the corresponding 2D arrangement with 4 molecules in the unit cell by alternating the chains F1 and F6, and relaxed it using our DFT method. The relaxed structure and its geometrical characteristics are shown in Fig. 3.18. One can recognise both types of chains there. Namely, looking from the top, we see F6-F1-F6-F1 sequence in the picture. However, the chains are distorted from their original geometry, as seen in Fig. 3.13. This

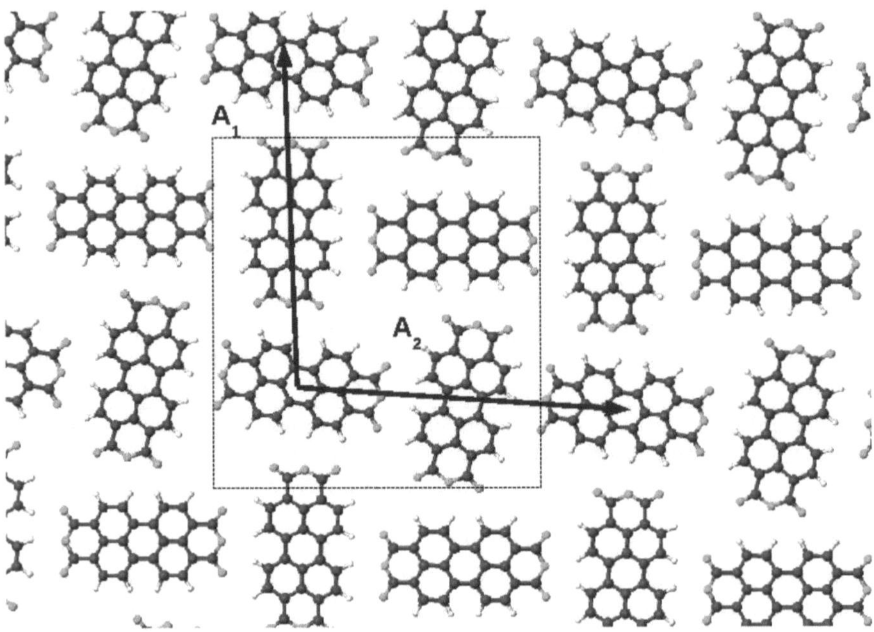

Fig. 3.18 The DFT relaxed geometry of the hexagonal phase based on the PTCDA tetramer unit cell. Both the lattice vectors, $\mathbf{A}_1$ and $\mathbf{A}_2$, and the unit cell are indicated ($|\mathbf{A}_1| = 23.51$ Å, $|\mathbf{A}_2| = 23.35$ Å, $\gamma = 83.6°$)

findings seem to indicate that the hexagonal structure seen in [17] may have been additionally stabilised by the interaction with the Si/Ag(111) surface. Note that the experimentally observed distance between equivalent molecules along the rows is of 23.03 Å [17] and it agrees well with the length of the lattice vector $\mathbf{A}_2$ found for the structure in our calculations.

Interestingly, the stabilisation energy for this structure found in our DFT calculations (−1.14 eV with respect to 2 molecules) differs very little from the stabilisation energies of other structures shown in Table 3.11. This result may explain why the three phases (herringbone, square and hexagonal) coexist on the Si/Ag(111) surface [17]. Similarly, we can rationalise the complex structure seen in the STM image of Fig. 1.4a [16]. In this image one can notice three regions: in the region on the right and on the left a herringbone phase is clearly visible, while in the region between them (as indicated by the arrow at the top) a narrow stripe of the square phase exists. Clearly, the two borders between the square and the herringbone phases are provided by exactly the same two rows F1 and F6 as in the hexagonal phase discussed above (which was, however, observed on a different Ag/Si(111) surface). Obviously, by alternating these two rows, the hexagonal phase appears. However, since one can repeat identical rows as well, as follows from our calculations on two molecule per cell structures as described in the previous section, these two chains provide a perfect arrangement for the herringbone phase to go continuously into the square phase and then back to the herringbone one. The particular key to the sequence seen in the STM image may be assigned as ...-F6-F6-F1-F1-F1-F6-F6-... assuming that the square type row F1 is repeated exactly three times. This sequence is approximate due to the relaxation of the molecules at the boundaries. Each particular arrangement is a result of complicated kinetics governing the formation of the assemblies during the deposition and (possible) further treatment.

Using our theoretical method, we find a large number of 1D and 2D structures for PTCDA molecules. Our calculations show that the chains and the monolayers based on PTCDA dimers are very similar in energy and in many cases also in geometry. However, one of the most stable PTCDA monolayers was found to be MON4, shown in Fig. 3.14, where the geometry of the system allows for the formation of more hydrogen bonds. Three of the obtained monolayers have been already observed experimentally as the square [17–20, 22, 24, 26], herringbone [17, 19, 20, 25–28] and brick wall [24, 29] phases. Some other structures we find theoretically have not been observed yet. In fact, they may never be identified due to their similarity to the already observed structures; at the same time, some of the structures have relatively small stabilisation energies and hence the probability of their formation is low.

In summary, a new complex structure was formed by the molecules on the Au(111) surface consisting of a herringbone phase continuously merging with the square phase; when the latter goes over into the herringbone phase again on the other side, the two herringbone phases appear to be shifted with respect to each other. That shift of course would depend on the width (i.e. the number of rows) of the intermediate square phase separating the two herringbone phases. Our theoretical analysis does not only help to rationalise this structure (as well as the previously observed on the Ag/Si(111) surface hexagonal phase), but demonstrates that such a boundary between the two distinct phases is *not* energetically prohibitive.

A detailed comparison between theory and experiment is hampered by a number of factors. Many experimental studies reported incommensurability of the molecular structures with respect to the periodicity of the substrates leading in some cases to very large cell extensions needed to match the lattice vectors of the surface (see, e.g. [17, 22]). Moreover, experimental observations reveal that the geometries of the same types of structures differ depending on the surfaces, e.g. various herringbone phases are seen on the (111) terminations of Cu [22], Au [23] and Ag [24]. The effect of the surface may be significant in some cases discussed above (e.g. for the hexagonal phase discussed in Sect. 3.2.4) and hence may render some of the structures we find theoretically to be unfavourable on a given substrate. These factors hint to the role played by surfaces in providing a platform for the arrangement of the molecules. A direct comparison of calculations and experimental STM (AFM) images would therefore require taking into account the interaction of the molecules with the surface directly, i.e. going beyond the gas-phase model adopted here. This is made even more difficult by the problem of incommensurability mentioned above and inability of commonly used density functionals to account for the dispersion interaction which may play a significant role in binding molecules on some surfaces such as e.g. the Au(111) surface [30].

Table 3.14 Stabilisation energies (eV) of the PTCDI dimers

	D1	D2	D3	D4	D5
E_{stab}	−0.15	−0.24	−0.5	−0.11	−0.14
E_{def}	0.05	0.07	0.11	0.03	0.01
E_{int}	−0.20	−0.31	−0.61	−0.14	−0.15
E_{BSSE}	0.14	0.14	0.15	0.12	0.09

The stabilisation and interaction energies include the BSSE correction which is also shown

3.3 PTCDI

The structure of the PTCDI molecule is close to that of the PTCDA, instead of the central oxygen atom at shorter sides there is an imide group, see Fig. 3.11. This small change in the molecule design leads, as we shall see, to essential differences in structures these molecules can form with each other. Although some of the structures are similar to those formed by PTCDA [11], there are also very different monolayers possible which are entirely due to an ability of the PTCDI molecules to form dimers which are unique to them.

In this section we shall consider all possible gas-phase structures based on two PTCDI molecules in the unit cell. We will describe in detail the approach used presenting first the PTCDI dimers the section that follows concern the building of all superstructures based on the dimer unit cells.

3.3.1 *PTCDI Dimers*

Again, we start by identifying the binding sites of the PTCDI molecule, see Fig. 3.11. Five possible types of binding sites can be identified, where three of them are exclusively donors (sites 2, 3 and 4) composed of carbon-hydrogen groups and are able to form double or triple hydrogen bonds with sites presenting acceptor atoms. In the case of PTCDI such purely acceptor sites do not exist (but they exist in the case of PTCDA [11]), so that sites 2, 3 and 4 cannot be functional. However, as we shall see in the following, two of the possible dimers do involve these sites to some extent. The two other types of binding sites have both donor and acceptor atoms (sites 1 and 5) composed of oxygen and hydrogen atoms which are able to form a double hydrogen bond with a binding site of the same type (reversed). There are four binding sites 1 and 5 around the molecule perimeter.

Combining binding sites 1 and 5, it is possible to obtain three pairs, D1, D2 and D3, whose DFT relaxed structures (Table 3.14) are shown in Fig. 3.19. In each case the dimers are stabilised by two hydrogen bonds. We also find that two other pairs are stable, D4 and D5, shown in Fig. 3.19. We tried these combinations inspired by some experimental observations [31]. In these two pairs purely hydrogen sites (2, 3 and 4) are also employed in some ways.

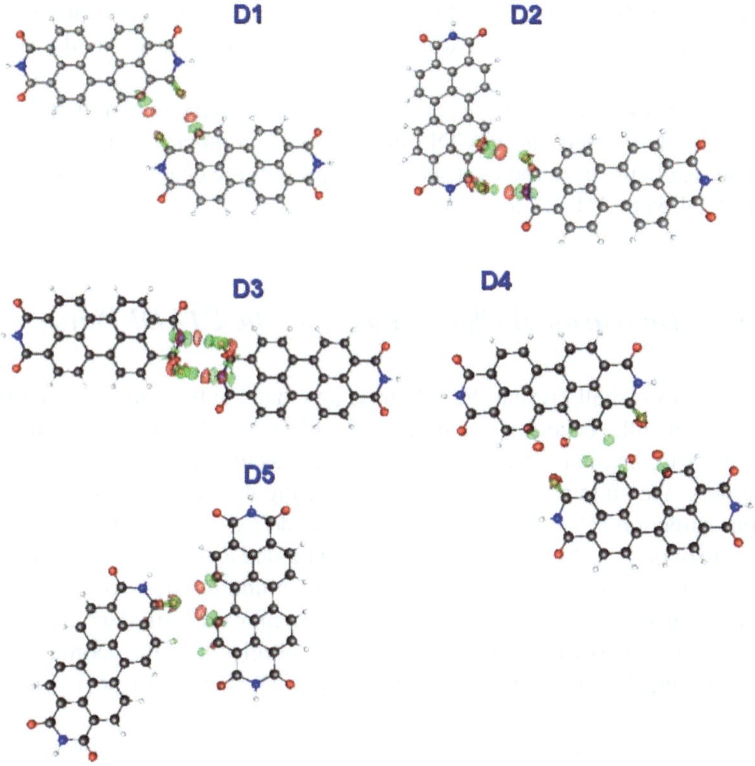

Fig. 3.19 The relaxed geometries of the PTCDI dimers shown together with the corresponding electron density difference plots corresponding to ± 0.01 electrons/Å^3. The *green* surfaces correspond to the regions of positive electron difference (excess) and the *red* areas correspond to the regions of negative electrons density difference (depletion)

The most stable dimer is D3 (see Table 3.14), and it is stabilised by a double hydrogen bond between the imine groups of the two PTCDI molecules. The two PTCDI molecules are connected in a "head-and-tail" structure. The dimer D1 is similar to the dimer D3 of PTCDA [11], however, the stabilisation energy in the case of PTCDI is lower. This difference in energies is attributed to the difference in the electronic structure of the two molecules. The dimer D2 has an L-shape and a much lower stability because one of the hydrogen bonds (out of two, see Fig. 3.19) is formed by the C–H group of one molecule with the oxygen of another. Very weak dimers D4 and D5 (see Table 3.14) are stabilised by hydrogen bonds formed between the oxygen atoms of the two PTCDI molecules and the hydrogen along the long side of the molecules. Interestingly, the dimer D5 is stabilised only through a little bit more than a single hydrogen bond (see Fig. 3.19) and has a stabilisation energy close to that of D1. one should recall that the dimers D4 and D5 have been identified in some of the experimental images [31]. As in the case of PTCDA [32], when carbon atoms are involved in the hydrogen bonding (D1, D2 and D5 dimers), the stabilisation energies are very small and the dimers are weak. Deformation energies are in all cases very small.

Once all the possible dimers are constructed, the next step is to identify the binding sites available to connect the PTCDI dimers with each other in such a way so that a 1D chain is obtained. In the next section we present the chains based on the most stable PTCDI pairs, D1, D2 and D3.

3.3.2 One-Dimensional Chains Based on the PTCDI Pairs

In this section we shall present the 1D structures that can be built using the PTCDI dimer as a unit cell. As before, we start by identifying the binding sites available on the periphery of the dimer, see Fig. 3.19, which are available for the connection with more dimers. We build the 1D chains by connecting dimers in a chain-like manner along one direction. The chains are obtained by adding a second dimer to a certain binding site of the first dimer, then a third dimer is added to the same site of the second one, and so on. In this way, using the available dimer binding sites, 16 different chains can be constructed which can be classified into families of "similar" chains, based on the evaluated energy (via summing up dimer energies) and the geometry of each chain. Following this procedure, six families of PTCDI chains have been constructed, see Fig. 3.20.

We present here, with a single exception, only the evaluated stabilisation energies, based on the sum of the corresponding dimer energies, without performing an additional relaxation of the structures using the DFT method. This is because we have found, in the case of PTCDA chains, that the evaluated and DFT calculated stabilisation energies of the chains are very close [16]. In the PTCDI case the same behavior is expected.

As it might be expected, the weakest chain is F3 that involves one of the weakest dimer contacts D1, while the most stable chain is F4 that involves by far the strongest dimer D3, see Table 3.15. One particular chain from this family, F4a, was observed on the Ag terminated Si(111) surface [17], and that is why we relaxed this chain with our DFT method. A favourable agreement of its geometry can be found with the one observed in Ref. [17], as is evidenced from Table 3.16. The binding energy for the chain was found to be -1.12 eV (the BSSE correction 0.34 eV) which is only slightly higher than the evaluated energy of -1.0 eV reported in Table 3.15. We expect the strength of the F4b chain to be similar or even identical to that of F4a.

3.3.3 Monolayers Based on the PTCDI Dimers

To build all possible 2D structures involving the PTCDI molecules, we followed the same procedure as in the cases of Melamine [10] and PTCDA [16]. Although

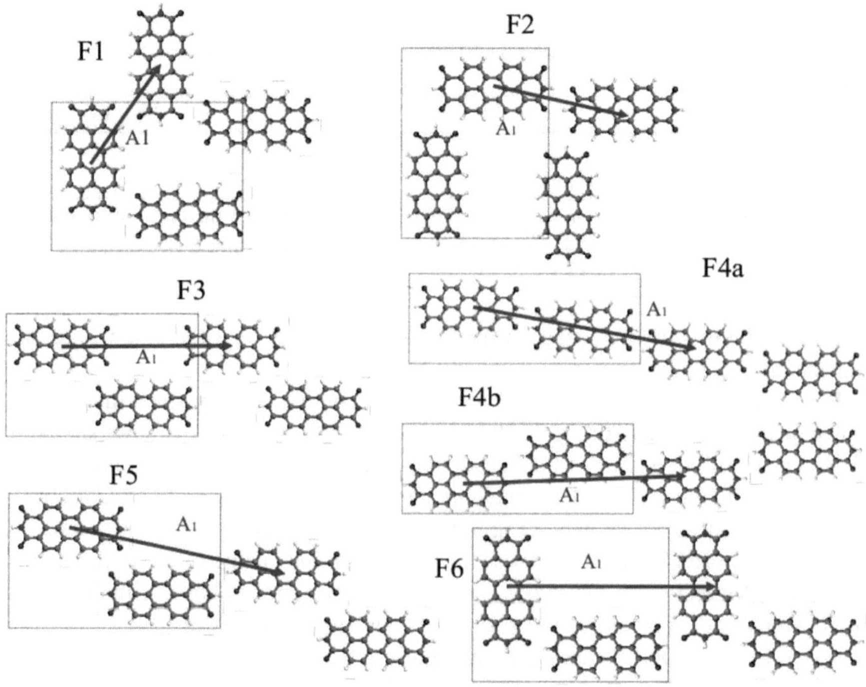

Fig. 3.20 Predicted structures for the selected six PTCDI chains with two molecule in the unit cells shown explicitly. To facilitate comparison with available experimental data, two chains from the same family F4 are shown here, designated as F4a and F4b. The structure of the F4b chain shown is the only one obtained after the full DFT relaxation. The lattice vector, $\mathbf{A}_1$, and the unit cell are indicated in each case

Table 3.15 The evaluated (by summing corresponding dimer energies) stabilisation energies (in eV) for selected PTCDI chains

	F1	F2	F3	F4	F5	F6
E_{stab} (eval.)	−0.39	−0.74	−0.30	−1.00	−0.65	−0.48

there are 16 1D possible chains (see Sect. 3.3.2) based on the strongest dimers D1, D2 and D3 considered above, due to the high symmetry of the molecule, only 11 2D structures can be constructed with two molecules in the unit cell using the same dimer connections. Once the monolayer energies are evaluated, it is possible to classify the structures into families of "similar" structures taking into account the geometries and the evaluated energies. In this way, seven families of monolayers can be identified and one representative from each was considered as a prototype. The selected seven monolayers are shown in Fig. 3.21, their energies are given in Table 3.17, while the geometrical characteristics of the unit cells in Table 3.18. One

Table 3.16 Comparison of the geometrical characteristics of the (canted) F4a chain (with one molecule in the unit cell), relaxed with our DFT method, with the one observed by STM on the Ag/Si(111) surface [17]

	STM [17]	This work (DFT)
A_1	14.1 ± 0.2	14.52
β	8 ± 2	10.4

A_1 is the length (in Å^3) of the lattice vector (i.e. the distance between the molecules) and β (in degrees) is the canting angle (the angle molecules make with the direction of the chain)

more monolayer MON8, based on the D4 and D5 dimers (or the chain F4a), was considered separately.

Note that the MON5 and MON7 monolayers contain, strictly speaking, just one molecule in the primitive unit cell. However, to simplify our discussion, we shall refer to them as containing two molecules, i.e. the corresponding supercells (as defined in Fig. 3.21) will be considered as an elementary unit. The structure MON2 is similar to the domino structure shown in Fig. 1.10b [32]. The structure MON1 is of a similar nature, more porous, and we shall use the same "domino" name for it. The structures MON3 and MON4 represent two variants of the herringbone phase, with the MON3 monolayer containing elongated pores between molecules, while MON4 is more compact. In the MON5 structure we recognise the brick wall phase observed in the STM images of [32] and shown in Fig. 1.10c. MON6 is a somehow similar phase with large pores formed by six molecules, and we coined the name "porous" for it. In MON7, molecules in each horizontal row go in a wave-like manner, so that we shall call this monolayer "waveform". Finally, MON8 must be the canted structure seen previously [17, 32–34], see Fig. 1.10a.

The most stable PTCDI monolayer is the canted structure MON8, see Table 3.17, based on the most stable chain F4a from the previous section. Moreover, this monolayer is also more stable than any of the PTCDA monolayers we have considered in Ref. [16] which is not surprising since the PTCDI molecules form a dimer (D3) by far more stable than any of the PTCDA dimers. The calculated canted angle (see the previous section) along the F4a rows in the relaxed structure is $\beta = 11.5°$. In the present case the evaluated stabilisation energies of monolayers are all smaller than the DFT calculated energies indicating the existence of cooperative effects.

The electron density plots for four of the calculated monolayers are shown in Fig. 3.22. One can clearly see the formation of hydrogen bonds of various strength between the molecules. The monolayer MON8 (canted phase) shows well developed hydrogen bonds along the F4a chains. In spite of the fact that the chains themselves are not bound together by the strongest D5 dimer connections (see Table 3.14), this monolayer is the strongest one due to favourable positioning of the molecules such that their binding along the chains is not severely affected. The MON7 monolayer also demonstrates quite developed "kebab" structures along the chains, however, relatively weak hydrogen bonds between the chains (made of D1 dimer connections, Table 3.14) yield this monolayer to be only next to the strongest one. This monolayer is then followed in energy by MON2 (the domino phase) where each molecule is

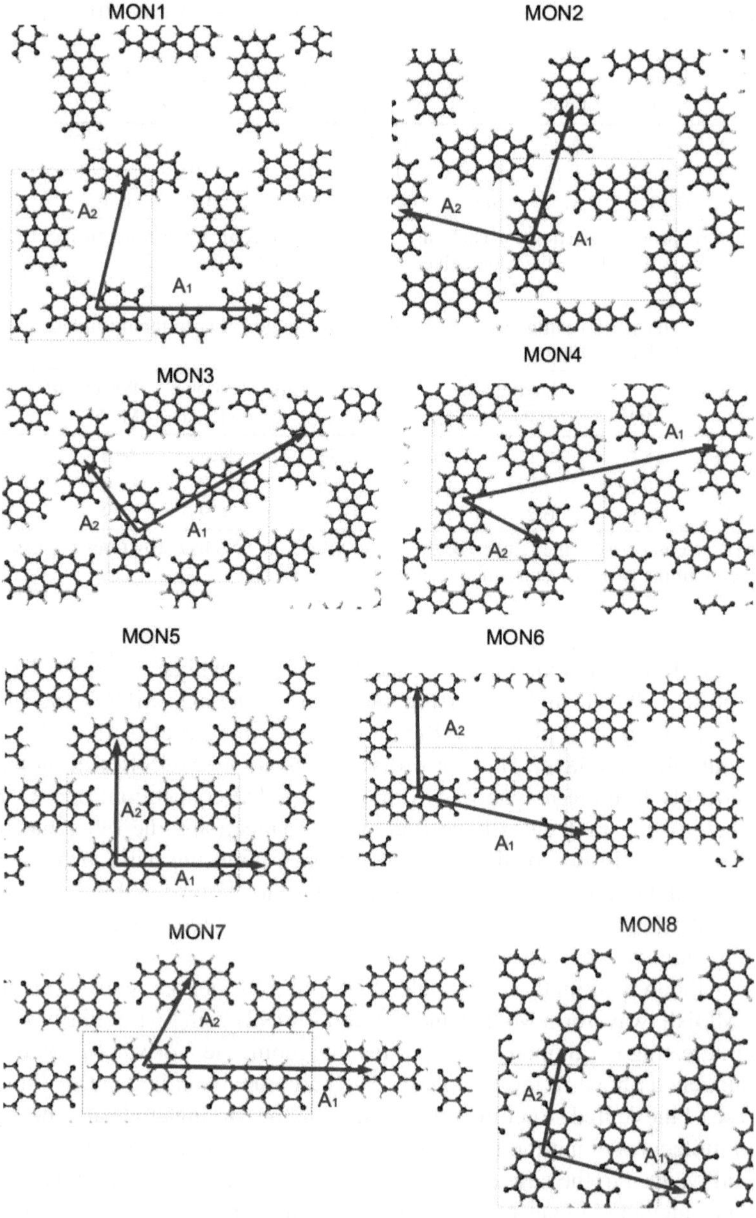

Fig. 3.21 DFT relaxed geometries of the selected eight prototype PTCDI monolayers. Lattice vectors as well as the unit cells are indicated. In the cases of MON5 and MON7 structures, based on primitive cells with one molecule, the lattice vectors shown correspond to a supercell containing two molecules

Table 3.17 The evaluated and calculated stabilisation energies (in eV) for PTCDI monolayers shown in Fig. 3.21

	Domino		Herringbone		Brick wall	Porous	Waveform	Canted
	MON1	MON2	MON3	MON4	MON5	MON6	MON7	MON8
E_{stab} (eval.)	−0.72	−0.96	−0.63	−0.72	−0.60	−0.80	−1.30	−1.55
E_{stab} (calc.)	−0.89	−1.12	−0.97	−0.91	−0.85	−0.99	−1.38	−1.67
E_{BSSE}	0.43	0.57	0.43	0.49	0.48	0.40	0.46	0.67

The BSSE correction is also shown in each case. For the ease of comparison, energies for the MON5 and MON7 structures refer to the supercell containing two molecules

Table 3.18 The lengths of lattice vectors (in Å) and the angle between them (in degrees) for the eight selected PTCDI monolayers

	Domino		Herringbone		Brick wall	Porous	Waveform	Canted
	MON1	MON2	MON3	MON4	MON5	MON6	MON7	MON8
A_1	22.25	18.0	26.11	26.12	18.84	24.3	28.63	17.45
A_2	17.92	18.01	12.05	12.23	15.86	15.9	11.62	14.51
γ	75.3	89.9	96.2	118.0	90.0	76.9	62.9	91.4

For the ease of comparison, the lattice vectors of the MON5 and MON7 structures correspond to a supercell containing two molecules in agreement with Fig. 3.21

connected to four others via the D2 connection of intermediate strength. The other domino phase (MON1) and the herringbone phase structures (MON3 and MON4), as well as the brick wall phase (MON5), and the porous structure (MON6) are the least favourable monolayers, but for different reasons: in MON1 each molecule is connected only to three neighbors via rather weak D2 connection; in MON3 and MON4 half of the molecules have four weak connections, the other half of the molecules have only two of them; in MON5 each molecule has four neighbours, however, the dimer connections (which is of the weak D1 type, Table 3.14) between them is weak; finally, in MON6 each molecule has only three connections including one strong and two weak ones.

So, our theoretical method has yielded six different phases, with three of them (canted MON8, waveform MON7 and domino MON2) being particularly stable.

It is interesting to compare the experimental geometries for various phases with those calculated above in the gas phase. Such a comparison is done in Table 3.19. In the case of the canted structure, one can see that the canting angle and the distance between molecules along the F4a row do not depend significantly on the substrate (surprisingly, even in the case of a not-so-flat Au(11,12,12) surface) and are similar to our data calculated in gas phase. However, other characteristics of the monolayer associated with the relative positions of the rows with respect to each other (A_1 and the angle γ), may noticeably depend on the substrate, although experimental data on these structures are only available for the Au(111) surface. This behaviour is explained by extremely strong interaction between molecules along the F4a row as compared with their interaction between the rows.

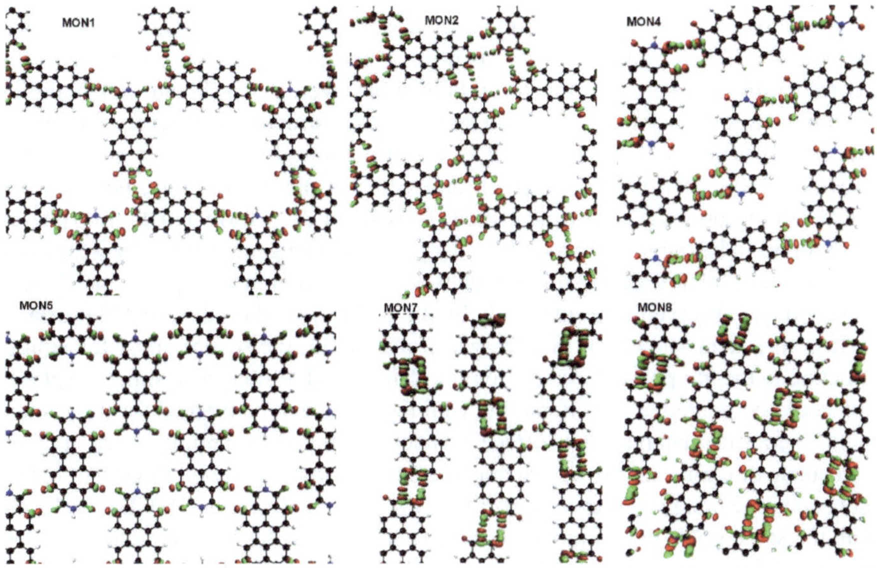

Fig. 3.22 The electron density difference plots corresponding to ±0.01 electrons/Å^3 of four PTCDI monolayers MON1, MON4, MON7 and MON8. The *green* surface corresponds to the regions of positive electron difference (excess) and the *red* areas correspond to the regions of negative electrons density difference (depletion)

Table 3.19 The lengths (in Å) of the two lattice vectors and the angle γ (in degrees) between them for the observed experimental image of the canted PTCDI monolayer

	Surface	References	A_1	A_2	γ	β
Canted	Au(11,12,12)	[34]	–	14.2 ± 0.2	–	11 ± 2
	Ag/Si(111)	[17]	–	14.6 ± 0.2	–	–
	Au(111)	[33]	19.8	14.5	75	12
		[32]	16.7	13.8	80	12
		[32]	17.45	14.51	91.4	11.5
Brick wall	Au(111)	[32]	18.7	17.0	90	–
			18.84	15.86	90	–
Domino	Au(111)	[32]	17.9	17.0	85	–
			18.0	18.01	89.9	–

β is the canted angle along the F4a type rows (in degrees) for the canted structure

The agreement between the gas phase calculations and the observed structures on gold for the brick wall and domino phases is much better: we observe very close unit cells in each case.

The forces stabilising the PTCDA and PTCDI structures are the same that stabilise smaller structures based on NTCDA and NTCDI molecules.

We have studied the periodic assemblies of PTCDI molecules may form on the Au(111) surface.

We followed a systematic theoretical approach in constructing 1D and 2D planar periodic PTCDI assemblies, as it was done in the previous sections. In this way, a considerable number of monolayers is built, which were then split into eight families of similar structures. After choosing a representative structure from every family, we proceeded further and relaxed the selected structures. In this way eight possible structures have been predicted which represent six phases: domino, herringbone, brick wall, porous, waveform and canted. The canted, waveform and domino phases were found to be the most favourable energetically. The geometrical characteristics of the canted, brick wall and domino phases calculated in the gas phase match very closely those observed experimentally by the STM method on the Au(111) surface.

It is surprising, however, that the waveform phase, the second most favourable structure in the gas phase, has not been found in our STM images (a simple explanation could be related to the interaction with the surface to be blamed for this); since in our theoretical analysis the surface was not accounted for and, if it was, it could have made this particular structure much less favourable. However, in view of recent more accurate first principles calculations [30] of a (quite similar) PTCDA molecule with the same gold surface in which the dispersion (van der Waals) interaction was accounted for within the newly developed vdW-DF functional [35, 36], this explanation seems to be unlikely since these calculations demonstrated quite clearly that the molecule-surface interaction is rather flat laterally. Therefore, this point still remains unclear and awaiting explanation.

There is also another point worth mentioning, in the case of the PTCDI molecules, one dimer connection is two times stronger than the best PTCDA dimer [16]. Nevertheless, the best 2D structure we find (the canted monolayer) has a only slightly higher binding energy when compared to the best PTCDA monolayer, which is unexpected. This result clearly demonstrates the importance of other dimer connections involved in the binding of the monolayer: sometimes, in order to accommodate the strong bondings, weak connections must also be used, and this may reduce the overall stability of the assembly.

3.4 NTCDA

In this section we shall consider our theoretical method in application to all possible gas-phase periodic structures formed by NTCDA molecules. The geometrical properties of NTCDA are the similar to those of the PTCDA, however, the NTCDA molecule is shorter.

3.4.1 NTCDA Dimers

The starting point in our analysis is the identification of the binding sites in the NTCDA molecule, see Fig. 3.23. Three possible non-equivalent binding sites have

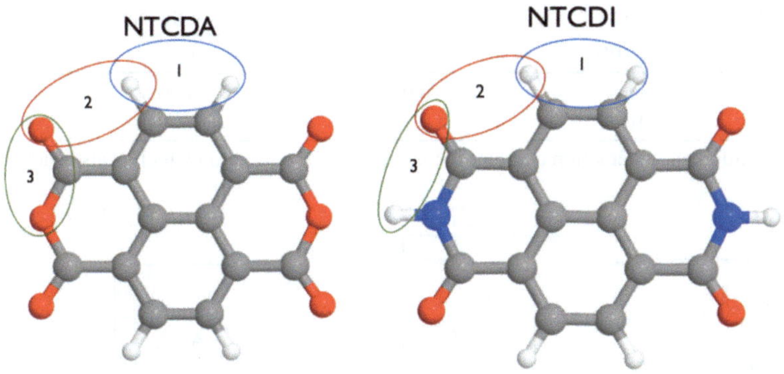

Fig. 3.23 NTCDA and NTCDI molecules and their non-equivalent binding sites shown by *numbered ovals*

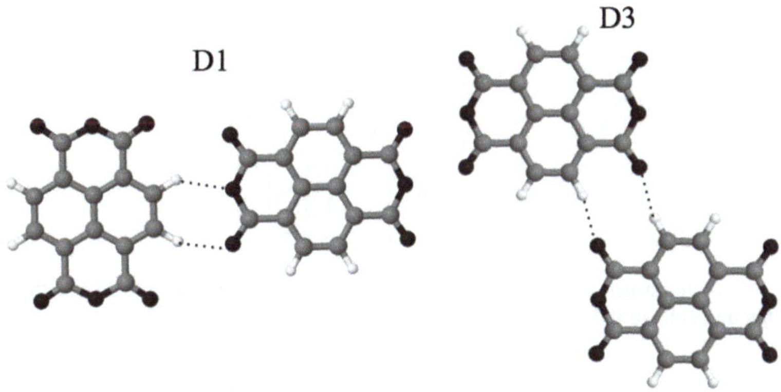

Fig. 3.24 The geometries of the two possible NTCDA dimers

been identified. Sites 1 are of the donor type: they are composed of carbon-hydrogen groups and are able to form double hydrogen bonds. Two of the binding sites are exclusively acceptors (3) and are composed of oxygen atoms enabling them to form two or three hydrogen bonds. There is also a third kind of binding site (site 2) with both donor and acceptor atoms (it is composed of an oxygen and hydrogen atoms able to form a double hydrogen bond); there are four such bonds along the molecule perimeter, similarly to PTCDA.

Although, the NTCDA molecule has few, both different and similar, binding sites, because of the symmetry of the molecule, only two dimers are possible and they are shown in Fig. 3.24. One of the dimers is constructed by placing one NTCDA molecule perpendicularly with respect to the other and shifting it along its longer side. Dimer D3 is obtained when the two NTCDA molecules are connected in a "head-and-tail" like structure, see Fig. 3.24, using binding site 3.

Table 3.20 Stabilisation (eV), deformation and interaction energies of the NTCDA dimers

	E_{stab} (PBE)	E(BSSE)	E_{def}	E_{int}
D1	−0.06	0.06	0.002	−0.06
D3	−0.16	0.13	0.004	−0.16

The stabilisation and interaction energies include the BSSE correction which is also shown

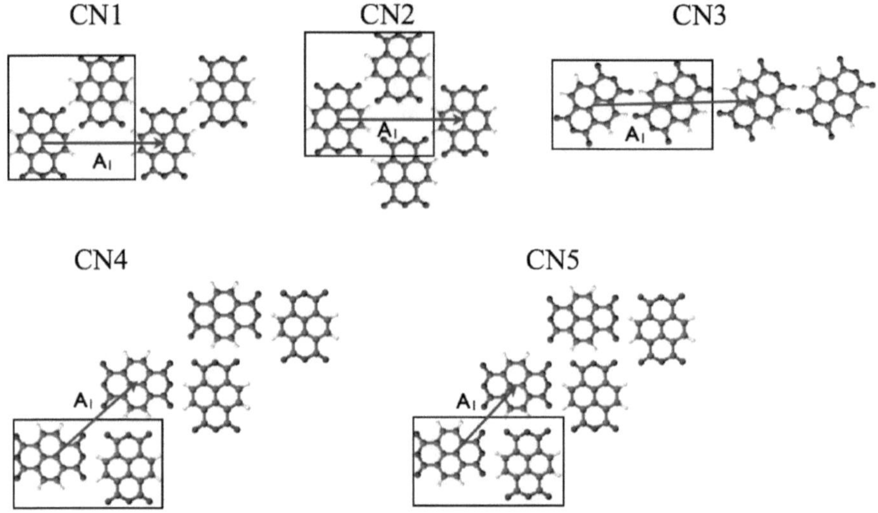

Fig. 3.25 Predicted structures for the selected five chains of NTCDA molecules. The lattice vector, $\mathbf{A}_1$, and the unit cell are indicated in each case

The two dimers are found to relax into stable geometries. As expected, the stabilisation energies of the dimers shown in Table 3.20, are found to be rather small and similar.

As evidenced from the binding energies, in all cases the dimers are very weak. By a small margin, the most stable dimer is D3. The properties of those dimers are very similar to PTCDA dimers, due to the similar chemical structures of PTCDA and NTCDA molecules.

3.4.2 One-Dimensional Chain Based on the NTCDA Dimers

In this section, following the same method used in the previous sections we shall present the 1D structures that can be built using the NTCDA dimer as a unit cell. We build the 1D chains by connecting dimers in a chain-like manner along one direction. Using the available dimer binding sites, 5 different chains can be obtained (Fig. 3.25).

The weakest chains are CN4 and CN5 that involve the weakest dimer contact D1, while the most stable chains are CN1, CN2 and CN3 that involve the strongest

Table 3.21 The evaluated stabilisation energies, in eV, (obtained by summing up the corresponding dimer energies) for NTCDA chains

	CN1	CN2	CN3	CN4	CN5
E_{stab} (eval.)	−0.32	−0.32	−0.32	−0.12	−0.12

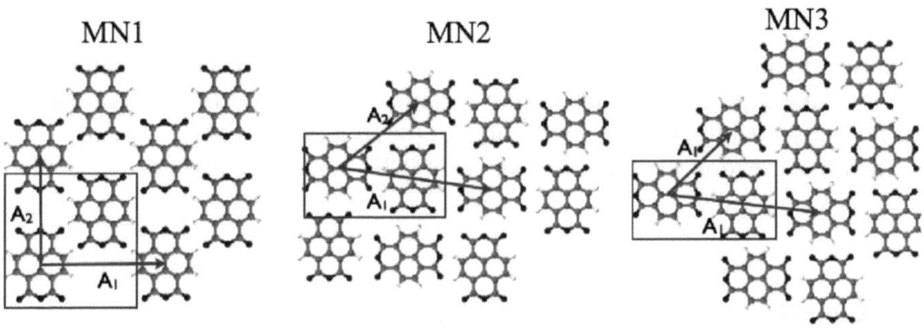

Fig. 3.26 Predicted structures for the three NTCDA monolayers with two molecules per unit cell shown explicitly. The lattice vectors and the unit cell are indicated in each case

dimer D3, see Table 3.21. However, one can build the chains CN1, CN2 and CN3 by considering only one molecule per unit cell.

We present here the evaluated stabilisation energies, based on the sum of the corresponding dimer energies, without performing an additional relaxation of the structures using the DFT method. This is because we find that, in the case of the PTCDA chains evaluated and the DFT calculated stabilisation energies of the chains are very close [16] (see Sect. 3.2.2). In the NTCDA case the same behavior is expected (Table 3.21).

3.4.3 Monolayers Based on NTCDA Dimers

In order to build the possible structures involving the NTCDA molecules, we followed the same procedure as in the previous cases (see Sects. 3.1.1.3, 3.2.3 and 3.3.3). Although there are five 1D chains (see Sect. 3.4.2), only three 2D structures can be constructed out of them with two molecules in the unit cell. We classified the 2D structures considering the similarity with the PTCDA monolayers. In this way, the following structures are obtained: a brick-wall structure based on the dimer D3 (MN1 see Fig. 3.26), and two herringbone-like structures based on the dimer D1 (MN2 and MN3 see Fig. 3.26).

We present the evaluated stabilisation energies, based on the sum of the corresponding dimer energies, without performing an additional relaxation of the structures using our DFT method. The most stable 2D structure is the monolayer MN1 based on the dimer D3, see Table 3.22. The monolayers MN2 and MN3 are less stable

Table 3.22 The evaluated (by summing up the corresponding dimer energies) stabilisation energies (in eV) for NTCDA monolayers

	MN1	MN2	MN3
E_{stab} (eval.)	−0.48	−0.18	−0.18

because they are based on the less stable dimer D1. It is possible that the stabilisation energy of the MN3 monolayer is slightly different because of extra-interactions, that are not accounted for without a relaxation of the structure, could be present.

In the next section we shall consider the NTCDI molecule and the possible dimer structures.

3.5 NTCDI

The structure of the NTCDI molecule is close to that of PTCDI. In this section we shall consider all possible gas-phase dimer structures.

3.5.1 NTCDI Dimers

Again, we start by identifying the binding sites of the NTCDI molecule, see Fig. 3.23. Three possible types of binding sites can be identified, where one of them is exclusively a donor (site 1) composed of carbon-hydrogen groups able to form double hydrogen bonds with sites presenting acceptor atoms. In the case of NTCDI, as for PTCDI [32], such purely acceptor sites do not exist (but they exist in the case of PTCDA [16] and NTCDA), so therefore site 1 cannot be functional. The remaining types of binding sites have both donor and acceptor atoms (sites 2 and 3) composed of oxygen and hydrogen atoms which are able to form a double hydrogen bond with a binding site of the same type (reversed).

Combining binding sites 2 it is possible to obtain two pairs, D1 and D2 whose structures are shown in Fig. 3.27. In each case the dimers are stabilised by two hydrogen bonds.

These structures have not been relaxed. However, due to the similarity with the PTCDI dimer, we expect that the most stable dimer is the D2. This dimer has the same characteristics as the dimer D3 composed by PTCDI molecules, see Fig. 3.19.

The next section is dedicated to the mixed phases between melamine and PTCDI/PTCDA molecules.

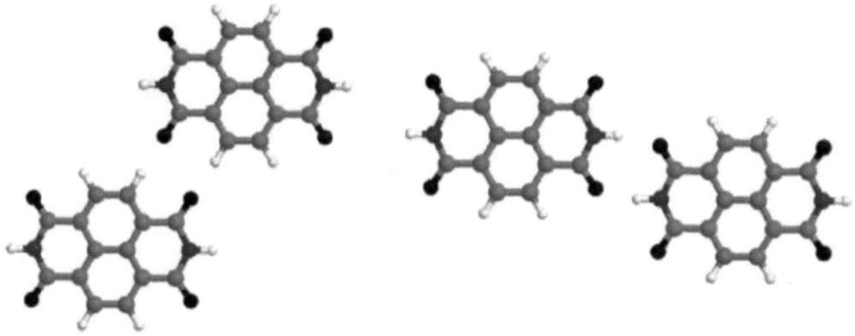

Fig. 3.27 The geometries of the two possible NTCDI dimers

3.6 Mixed PTCDA-Melamine and PTCDI-Melamine

It is well known [23, 37] that melamine and PTCDA (or PTCDI) molecules can form porous networks in which triangular melamine molecules serve as connectors. Here we shall consider the possible mixed networks which can be formed by these molecules. As done in the previous sections, the binding sites of each molecule have been identified so that all possible structures based on two or more molecules per unit cell have been considered and relaxed. The unit cell obtained combining either PTCDI or PTCDA and melamine are geometrically similar due to the similarity between PTCDA and PTCDI. In order to underline the difference between the mixed phases based on melamine-PTCDA and melamine-PTCDI unit cells, they will be presented at the same time. We shall describe in detail the approach used, as done on the previous structures, starting from the melamine-PTCDA/PTCDI dimer; this will then be followed by building all super-structures based on the dimer unit cell.

3.6.1 Melamine PTCDA/PTCDI Dimers

The starting point of our method is again the identification of the binding sites of the molecules as was done for melamine in Sect. 3.1.1, for PTCDA in Sect. 3.2.1, and for PTCDI in Sect. 3.3.1. Only two dimers are possible due to the geometrical properties of the molecules. The most stable of the two is the one obtained by connecting the side containing the oxygen atoms of the PTCDA molecules and one of the sides of a melamine molecule, forming a dimer stabilised by a double hydrogen bond, see dimer D1 in Fig. 3.28. The same kind of dimer can be constructed using the PTCDI and the melamine (see Di1 Fig. 3.28). Other possible dimers are obtained by connecting a hydrogen atom to one of the oxygen atoms at the corners of the PTCDA/PTCDI molecules and nitrogen and hydrogen atoms of melamine, in Fig. 3.28; In dimers D2 and Di2, the molecules are not in good alignment, thus resulting less stable.

The stabilisation energies of the PTCDA-melamine and PTCDI-melamine dimers, shown in Table 3.23, underline that dimer D1 and Di1 are very different in energy

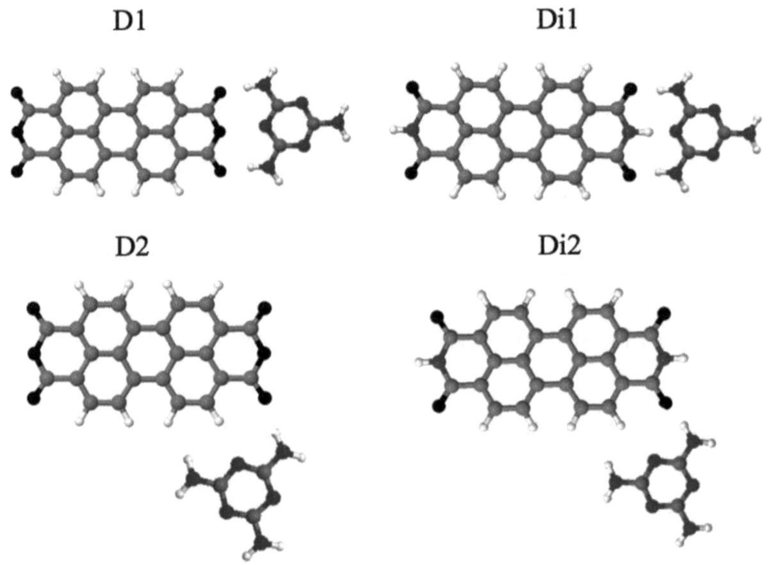

Fig. 3.28 Geometry of PTCDA/PTCDI melamine dimers

Table 3.23 Stabilisation, deformation and interaction energies (in eV) of the PTCDA-Melamine and PTCDI-Melamine dimers

	E_{stab} (PBE)	E(BSSE)	E_{def}	E_{int}
PTCDA-melamine (D1)	−0.18	0.14	0.01	−0.17
PTCDA-melamine (D2)	−0.32	0.13	0.01	−0.31
PTCDI-melamine (Di1)	−0.76	0.23	0.26	−0.50
PTCDI-melamine (Di2)	−0.30	0.13	0.05	−0.25

The stabilisation and interaction energies include the BSSE correction which is also shown

due to the difference in the structures of the PTCDA and PTCDI binding sites. In particular, the PTCDI-melamine Di1 structure is the most stable dimer while the D1 is less stable. The lower stability is due to repulsion between the central oxygen atom of PTCDA and the central nitrogen atom of melamine. In the case of PTCDI and melamine molecules, this repulsion is replaced by a hydrogen bond between the central N–H group of PTCDI and the nitrogen atom of melamine. On the other hand, the stabilisation energy of dimers D2 and Di2 are similar because the same kind of binding sites are involved and similar interactions are involved.

The dimers are the starting point to construct the 1D chains and the 2D monolayers that will be the subject of the next section.

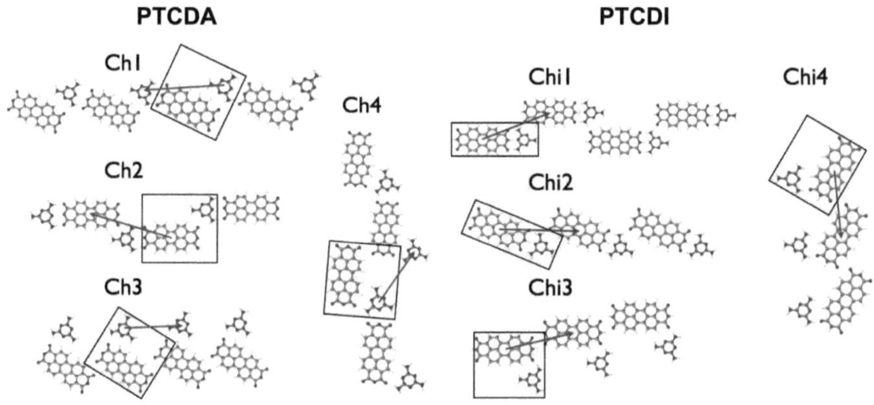

Fig. 3.29 Geometry of one-dimensional chains and the corresponding lattice vectors corresponding to a supercell based on the PTCDA-melamine and PTCDI melamine dimers

3.6.2 One-Dimensional Chains and Two-Dimensional Monolayers Based on Melamine-PTCDA and Melamine PTCDI Dimers

In this section, we will present 1D structures that can be built using the PTCDA/PTCDI-melamine dimers as a unit cell. As before, we start from the identification of the binding sites available on the periphery of the dimers.

By considering the PTCDI/PTCDA-melamine dimers, there are a large number of binding sites available to connect the dimers to each other. It is possible to obtain two 1D chains for each dimer, the total of four possible chains, see Fig. 3.29. Evaluation of the stabilisation energies of these chains is done as in the previous sections, by summing up the stabilisation energy due to single contacts.

As usual, by considering the contacts available in the chains, all the monolayers could be obtained. It is possible to construct only one monolayer based on the D2 dimer obtained by combining Ch1 and Ch2, see Fig. 3.30. In the same way, only one monolayer based on Di2 dimer is obtained, see Fig. 3.30. The two monolayers M1 and Mi1 are geometrically similar.

The evaluated stabilisation energy of the M1 monolayer is $E_{\text{stab}}(\text{eval.}) = -0.75$ eV, taking into account the D1 and D2 contacts between melamine and PTCDA molecules and the PTCDA-PTCDA contact, see Table 3.9. In the case of PTCDI, the evaluated stabilisation energy of the Mi1 monolayer is $E_{\text{stab}}(\text{eval.}) = -1.36$ eV. Since these 2D monolayers based on the dimer unit cell do not explain the STM images observed experimentally [1] we only evaluated the stabilisation energies here. In order to have a complete understanding of experimental structures, one has to consider structures with more than two molecules per unit cell. In the next section we shall describe what kind of structures are expected if one uses more than two molecules per unit cell.

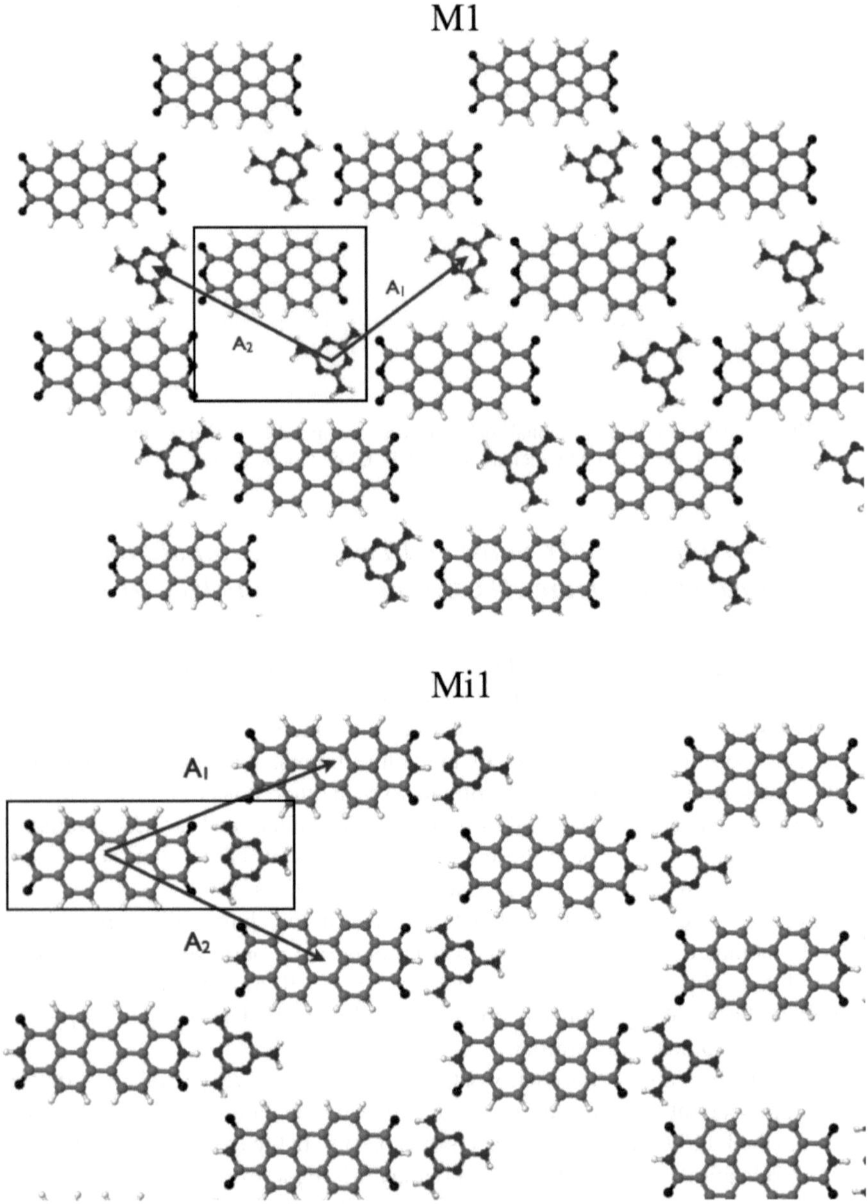

Fig. 3.30 Geometry of two-dimensional monolayers and the lattice vectors corresponding to a supercell based on the PTCDA-melamine D2 dimer and PTCDI-melamine Di2 dimer

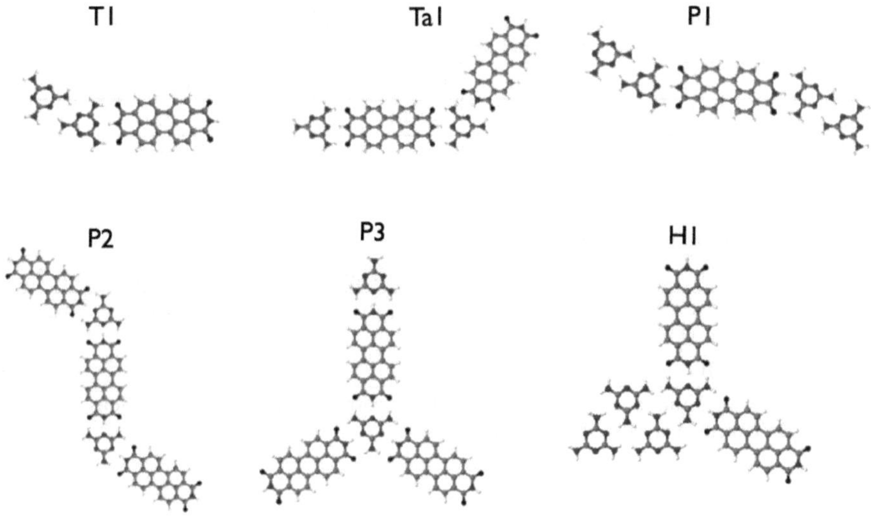

Fig. 3.31 Geometries of some of the PTCDI-melamine unit cells with more than two molecules

3.6.3 Going Beyond Two Molecules Per Unit Cell

The aim of this section is to explain the periodicity of some STM images observed by depositing PTCDI and melamine molecules on the Au(111) surface. These two molecules, once deposited on the surface, assemble and form interesting networks. Structures such as the parallelogram and hexagonal domains, see Fig. 1.12a, b, but also other mixed phases with different concentration of melamine and PTCDA/PTCDI, molecules see Fig. 1.12c, d, have been reported [1, 3, 37, 38]. The hexagonal phase exists alongside the parallelogram network. The structures presented in the previous section cannot describe the experiments because they have a different periodicity. We shall account for 2D structures based on more than two molecules per unit cell, for example three or four molecules, considering an alternation of PTCDA/PTCDI and melamine molecule on the unit cell.

The STM images described above and shown in Fig. 1.12 have been observed by depositing PTCDI and melamine on the Au(111) surface. In this section we only consider the assemblies which can be obtained combining PTCDI and melamine. In fact in the PTCDA-melamine case the connections are weak due to the reasons explained previously (Sect. 3.6.1) in addition to there being insufficient experimental work involving these molecules.

In order to consider monolayers that have been observed experimentally one can play with the ratio between the number of melamine and PTCDI molecules. In this way, five unit cells have been constructed: one trimer, one tetramer, three pentamers and one hexamer, all shown in Fig. 3.31. The trimer T1 is obtained combining two melamine and one PTCDI molecule, another possible trimer (T2) could be considered if two PTCDI molecules are combined with one melamine. The tetramer Ta1 is

Table 3.24 The evaluated (by summing up the corresponding dimer energies) stabilization energies (in eV) for selected PTCDI-melamine unit cell

	T1	Ta1	P1	P2	P3	H1
E_{stab} (eval.)	−1.20	−2.28	−2.40	−3.04	−3.04	−3.28

Table 3.25 The evaluated (by summing up the corresponding dimer energies) stabilisation energies (in eV) for selected PTCDI-melamine monolayers

	MON_T	MON_{Ta}	MON_{P1}	MON_{P2}	MON_{P3}	MON_{H1}	MON_{H2}
E_{stab} (eval.)	−2.4	−3.04	−3.28	−2.9	−4.56	−4.8	−4.8
E_{stab}	−1.77	–	−3.30	–	−4.44	−4.44	–
E_{BSSE}	0.81	–	1.07	–	1.36	1.38	–

The calculated stabilisation energies together with the corresponding BSSE corrections, obtained using our DFT method are also shown in some cases

obtained considering two PTCDI and two melamine molecules combined in such way that the molecules alternate; also in this case other tetramers can be considered. Three pentamers have been considered (P1, P2 and P3) and one hexamer H1, see Fig. 3.31. Combining these unit cells as done in the previous sections for the dimers, many 2D structures can be obtained, however, only a few of these are comparable with the experimental ones.

To give an idea of the stabilities of these structures, we evaluated the stabilisation energies of each unit cell considering the energy of every contact involved in the structure. The most stable structure is the hexamer H1 because of the larger number of molecules and hence connections between them. Unfortunately, it is impossible to compare the energies of these unit cells because different numbers of molecules are involved (Table 3.24).

Placing a larger number of molecules in the unit cell increases the number of networks that can be obtained and the possibility to have more porous patterns. In theory, a large number of monolayers could be considered, however, only few of them are comparable with those observed in experimental STM images and for our analysis we shall consider only the monolayers similar to those observed experimentally.

In Table 3.25 we show calculated (using our DFT method) and estimated (by summing up the appropriate dimer energies) stabilisation energies of some possible monolayers, shown in Fig. 3.32. One can see that, for the structures where the comparison is possible, the calculated energies are in good agreement with the evaluated ones.

By comparing the experimental images and our calculations, see Figs. 1.12 and 3.32, we can identify different networks. In order to distinguish the different phases we compared the geometrical properties of the networks on STM images and of the monolayers in Fig. 3.32. The parallelogram phase, see Fig. 1.12a, could be identified as the monolayer MON_{P2}, while the hexagonal phase, see Fig. 1.12b, could be identified as the monolayer MON_{P3}. The third and the fourth domains, see in Fig. 1.12c, are

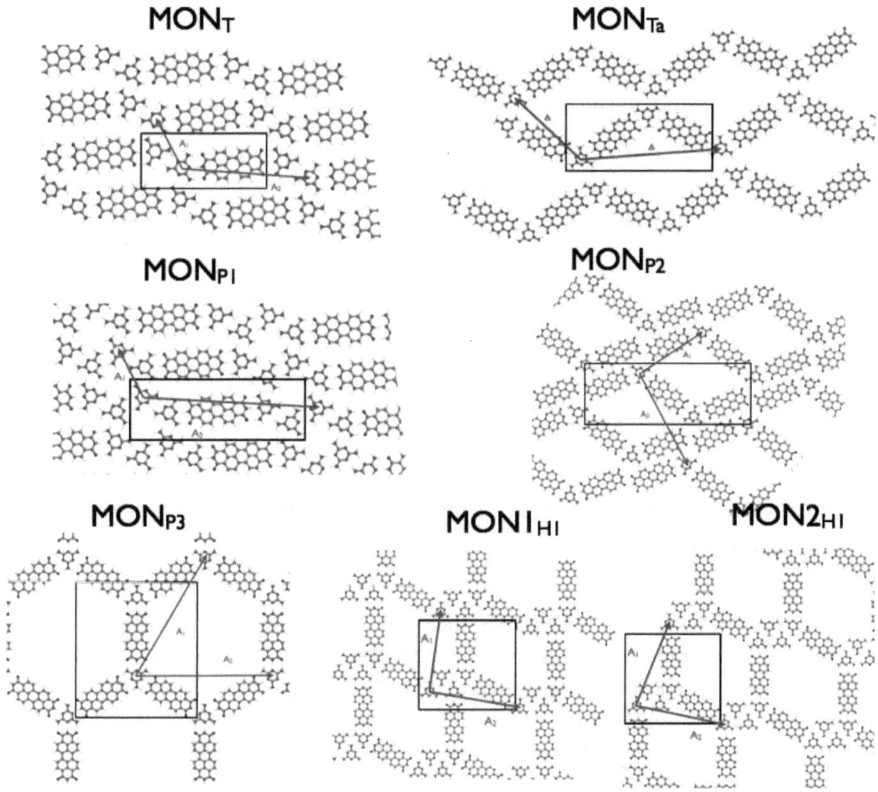

Fig. 3.32 Geometries of PTCDI-melamine monolayers based on unit cells with more than two molecules

the structures MON_T and MON_{P1}. These mixed phases can coexist when deposited on the surface as observed in Fig. 1.12d [23].

The geometrical characteristics of some of the structures, either relaxed with DFT (MON_T, MON_{P1} and MON_{P3}) or estimated from the dimer geometries (MON_{P2}), are compared with those observed in STM images in Table 3.26. The lattice vectors of the monolayers are in good agreement with those available in the literature.

Up to this point all our calculations have been performed using an approximation to the exchange-correlation energy in DFT that cannot describe the non-local dispersion forces. In order to take into account this type of interaction, we used a more accurate approximation to the exchange-correlation energy called the vdW-DF method (see Sect. 2.7). In the next section we analyse the role played by this non-local forces in the hydrogen bonded assemblies of molecules in the gas phase.

Table 3.26 The lengths of lattice vectors (in Å) and the angle between them γ (in degrees) for the selected PTCDI-melamine monolayers

		A_1	A_2	γ
Parallelogram	MON_{P2}	21.36	29.36	95
	[38]	19.4 ± 0.3	30.7 ± 0.4	–
Hexagonal	MON_{P3}	34.6	34.5	60
	[38]	35	35	60
	[37]	34.6	34.6	60
Mixed phase 1	MON_T	10.54	21.49	82.2
	[38]	–	21.9 ± 0.5	19.9
	[23]	10.0	19.9	85
Mixed phase 2	MON_{P1}	10.58	31.55	74.7
	[38]	–	31.5 ± 0.5	–

Theoretical and observed values are compared

3.7 Importance of vdW Interaction for Hydrogen Bonding Systems

In order to assess the role played by the vdW (or dispersion) interactions in the assembly of the molecules in the gas phase we firstly consistently performed geometry relaxation of the molecules themselves. Note that when doing the vdW-DF calculations, care should be taken in the choice of the cell dimension corresponding to the stacking direction of the molecules. We have checked that the total energies of our molecules change by less than 0.01 eV when this cell dimension changes from 20 to 40 Å, which is sufficient for our purposes. Our calculations confirmed that PBE and vdW-DF functionals give almost identical results for the bond lengths and bond angles which were found to agree within 0.01 Å and 0.1°, respectively. Then, relaxation calculations were performed for the corresponding flat melamine-melamine as well as PTCDA-PTCDA and NTCDA-NTCDA dimers. The latter were considered in two geometries D1 and D3 as shown in Fig. 3.33, and, similarly to the single molecules, no major differences were found in the relaxed geometries between the two density functionals.

As far as we are aware, no high-quality quantum chemistry calculations have been previously performed for these dimer systems. Therefore, in order to evaluate the performance of the two density functionals, binding energies of the dimers where also calculated using the MP2 method.[2] As demonstrated by Jureĉka et al. [8], it reliably describes hydrogen-bonded interactions. On the other hand, the benchmark calculations of Goll et al. [9] imply a decent performance of PBE for hydrogen-bonded complexes as well. The results of our dimer calculations are presented in Table 3.27. The dimers we considered are shown in Fig. 3.33. The binding energies obtained with the vdW-DF and MP2 are in reasonable agreement for all systems, however, the PBE energies for NTCDA and PTCDA dimers are too small. This is

[2] These calculation were done by Gulans.

Table 3.27 Stabilisation energies (eV) of the Melamine, NTCDA and PTCDA dimers, including PBE, vdW-DF and MP2 calculation

	PBE	vdW-DF	MP2
M–M	−0.48 [11]	−0.42	−0.49
NTCDA (D1)	−0.06	−0.18	−0.19
NTCDA (D3)	−0.16	−0.28	−0.28
PTCDA (D1)	−0.23	−0.38	−0.45
PTCDA (D3)	−0.25	−0.38	−0.35

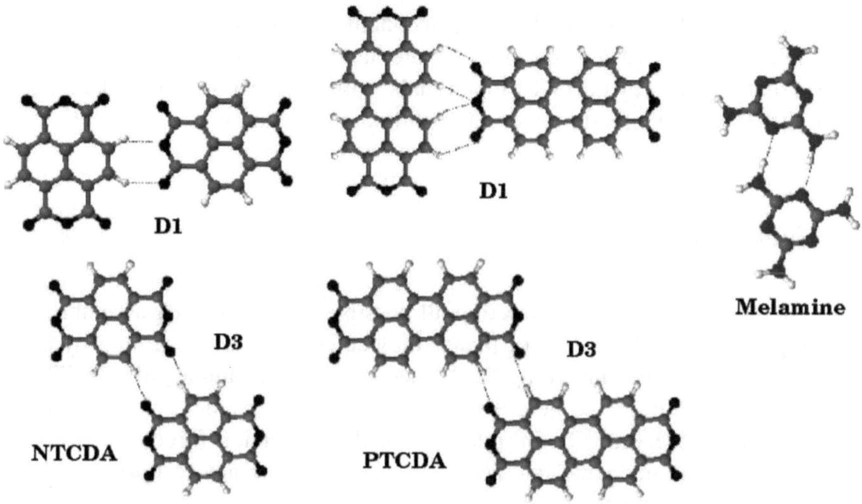

Fig. 3.33 Geometry of Melamine, NTCDA and PTCDA dimers

not surprising since it is well known that hydrogen bonds involving carbon atoms are very weak. What is surprising, however, is that in these cases there is a significant contribution of the dispersion interaction to the binding. On the other hand, in the case of the melamine dimer bound by strong double N–H...N hydrogen bonds, the dispersion interaction appears to be much less significant. This is also the smallest and therefore the least polarisable complex in our study. The binding obtained in PBE for the melamine dimer is in general accord with previous calculations of strong hydrogen bonded systems [39] which shows an insignificant effect of the dispersion interaction on this type of bonding as compared with the PBE calculations.

It follows from these results that vdW-DF functional performs extremely well when compared to the MP2 calculations. Not only is the order of stability in these systems is correctly reproduced, but also the absolute values of the binding energies are very close to the MP2 results. Note that, if instead of revPBE we had used the PBE exchange in our vdW-DF calculations, a somewhat better agreement with the

Coupled Cluster calculations could have been expected [35]. However, we did not exploit this avenue here as it goes beyond the scope of this study.

In the case of the strongly bound melamine dimer, we have also looked at charge density differences and found the characteristic "kebab" structure for the hydrogen bonding, which is almost identical to that obtained with the PBE density functional [10].

Weakly bound NTCDA and PTCDA dimers do not show any well developed "kebab" structures with either of the density functionals. These results may be extended to other molecules which may form hydrogen bonds with each other. Since the stability of two-dimensional assemblies bound by hydrogen bonding is, for the most part, due to the dimers involved [10, 40–43], we can conclude from these calculations that the vdW-DF method gives a description of the lateral interaction between molecules very close to that provided by the GGA functionals, with the exception of weak bonds for which the vdW-DF method provides a slightly higher binding.

3.8 Conclusions

Our systematic method of constructing all possible supramolecular assemblies has been applied to a wide variety of planar molecules capable of forming hydrogen bonding with each other, including DNA bases [42, 43], melamine [10, 11], PTCDA [16] and PTCDI, considered here. We obtained a good agreement with the patterns experimentally observed. The introduction of the dispersion forces when considering the molecule-molecule interaction was found to be not so important in the stabilisation of the network.

This study provides evidence that theoretical calculations performed in the gas phase are extremely useful, at least as a first approximation, for assessing the energetics and the arrangements of two-dimensional supramolecular architectures on surfaces where inter-molecular interactions dominate the molecule-surface interaction, or if the latter has a relatively small corrugation across the surface. The information on molecular energetics and possible arrangements is essential to tailor and engineer molecular assemblies at the molecular level in order to build sophisticated organic structures suitable for nanotechnology.

In spite of the promising results obtained in the gas phase, the next chapter is dedicated to investigating the role played by the surface in assembly of the molecules. Particular attention will be paid to characterise the potential energy surface PES of our molecules on the Au(111) surface.

References

1. Theobald JA, Oxtoby NS, Phillips MA, Champness NR, Beton PH (2003) Nature 424:1029
2. Knaggs EI, Lonsdale K, Wood RG, Williams G (1940) R. Soc. Lond. Proc. Series A 177:140–147

3. Perdigão LMA, Perkins EW, Ma J, Staniec PA, Rogers BL, Champness NR, Beton PH (2006) J Phys Chem B 110:12539
4. Staniec PA, Perdigão LMA, Rogers BL, Champness NR, Beton PH (2007) J Phys Chem C 111:886
5. Perdigão LMA, Champness NR, Beton PH (2006) Chem Commun 5:538
6. Ranganatha A, Pedireddi VR, Rao CNR (1999) J Am Chem Soc 121:1752
7. Zhang H, Xie Z, Long L, Zhong H, Zhao W, Mao B, Xu X, Zhao W (2008) J Phys Chem C 112:4209
8. Jurecka P, Sponer J, Cemý J, Hobza P (2009) Phys Rev B 79:201105
9. Goll E, Leininger T, Manby FR, Mitrushchenkov A, Werner H-J, Stoll H (2008) Phys Chem Chem Phys 10:3353–3357
10. Mura M, Martsinovich N, Kantorovich LN (2008) Nanotechnology 19:465704
11. Silly F, Shaw AQ, Castell MR, Briggs GAD, Mura M, Martsinovich N, Kantorovich LN (2008) J Phys Chem C 112:11476
12. Otero R, Shőck M, Molina LM, Lægsgaard E, Stensgaard I, Hammer B, Besenbacher F (2005) Angew Chem 44:2270
13. Kelly REA, Lee YJ, Kantorovich LN (2005) J Phys Chem B 109:22045
14. Kelly REA, Kantorovich LN (2006) J Mater Chem 16:1894
15. Kelly REA, Lee YJ, Kantorovich LN (2006) J Phys Chem B 110:2249
16. Mura M, Sun X, Jonkman HT, Silly F, Briggs GAD, Castell MR, Kantorovich L (2010) Experimental and theoretical analysis of h-bonded supramolecular assemblies of ptcda molecules on the au(111) surface. Phys Rev B 81(12):195412–195422
17. Swarbrick JC, Ma J, Theobald JA, Oxtoby NS, O'Shea JN, Champness NR, Beton PH (2005) J Phys Chem B 109:12167
18. Gabriel M, Stöhr M, Möller R (2002) Appl Phys A 74:303
19. Chinzhov I, Kahn A, Scoles G (2000) J Crys Growth 208:449
20. Mannsfeld S, Toerker M, Schmitz-Hubsch T, Sellam F, Fritz T, Leo K (2001) Org Electron 2:121
21. Kröger J, Jensen H, Berdt R, Rurali R, Lorente N (2007) Molecular orbital shift of perylenetetracaboxylic-dianhydride on gold. Chem Phys Lett 438:249–253
22. Wanger Th, Bannani A, Bobisch C, Karacuban H, Moller R (2007) Condens Matter 19:056009
23. Silly F, Weber UK, Shaw AQ, Burlakov VM, Castell MR, Briggs GAD, Pettifor DG (2008) Phys Rev B 77:201408
24. Glockler K, Seidel C, Sokolowski M, Umbach E, Bohringer M, Berndt R, Schneider W-D (1998) Surf Sci 405:1–20
25. Kunstmann T, Schlarb A, Fendrich M, Wangner Th, Moller R, Hoffmann R (2005) Phys Rev B 71:121403
26. Kilian L, Hauschild A, Temiroc R, Soubatch S, Scholl A, Reinert F, Lee T-L, Tautz FS, Sokolowski M, Ulbach E (2008) Phys Rev Lett 100:136103
27. Nicoara N, Romani E, Gomez-Rodriguez JM, Martin-Gago J, Mendez J (2006) Org Electron 7:287
28. Fendrich M, Kunstmann T, Paulkowski D, Möller R (2007) Nanotechnology 18:084004
29. Lauffer P, Emtsev KV, Graupner R, Seyller T, Ley L (2008) Phys Stat Sol B 245:2064
30. Mura M, Gulans A, Thonhauser T, Kantorovich L (2010) Role of van der waals interaction in forming molecule-metal junctions: flat organic molecules on the au(111) surface. Phys Chem Chem Phys 12(18):4759–4767
31. Perdigao LMA, Saywell GN, Fontes A, Staniec PA, Goretzki G, Phillips AG, Champness NR, Beton PH (2008) Chemistry (Eur J) 14:7600
32. Mura M, Silly F, Briggs GAD, Castell MR, Kantorovich L (2009) H-bonding supramolecular assemblies of ptcdi molecules on the au(111) surface. J Phys Chem C 113:21840–21848
33. Silly F, Shaw AQ, Briggs GAD (2008) Chem. Comm. 16:.
34. Cañas ME, Xiao W, Wasserfallen D, Müller K, Brune H, Barth JV (2007) Angew Chem Int Ed 46:1814–1818
35. Gulans A, Puska MJ, Nienminen RN (2009) Phys Rev B 79(20):201105–201108

36. Dion M, Rydberg H, Schroeder E, Langreth DC, Lundqvist BI (2004) Phys Rev Lett 92:246401
37. Swarbrick JC, Rogers BL, Champness NR, Beton PH (2006) J Phys Chem B 110:6110
38. Saywell A, Magnano G, Satterley CJ, Pedigão LMA, Champness NR, Beton PH, O'Shea JN (2008) J Phys Chem C 112(20):7706–7709
39. Cooper VR, Thonhauser T, Puzder A, Schroeder E, Lundqvist BI, Langreth DC (2008) J Am Chem Soc 130:1304
40. Kelly REA, Kantorovich LN (2005) Hexagonal adenine networks constructed from their homo-pairings. Surf Sci 589:139–152
41. Perdigão LMA, Staniec PA, Champness NR, Kelly REA, Kantorovich LN, Beton PH (2006) Adenine monolayers on ag-terminated si(111). Phys Rev B 73:195423
42. Kelly REA, Xu W, Lukas M, Otero R, Mura M, Lee YJ, Lægsgaard E, Stensgaard I, Kantorovich LN, Besenbacher F (2008) An investigation into the interactions between self-assembled adenine molecules and the au(111) surface. Small 4:1494
43. Kelly REA, Lukas M, Kantorovich LN, Otero R, Xu W, Mura M, Laesgaard E, Stensgaard I, Besenbacher F (2008) Understanding disorder of the dna base cytosine on the au(111) surface. J Chem Phys 129:184707

Chapter 4
Molecules on the Au(111) Surface

The previous chapter provided some evidence that calculations performed in the gas phase can be useful if the molecule-surface interaction has a relatively small corrugation across the surface. However, the presence of the surface can be important in giving a preferential growth direction or stabilising structures which are not the most stable in the gas phase. More complex structures may be possible, for which the surface may play some role, e.g. domain walls to relieve the strain between different domains. Therefore, the calculation of the binding energies of the molecules with the surface and the energy barriers for their diffusion are of crucial importance. In this chapter, we shall treat accurately these issues using a different kind of approach in order to have a complete characterisation of the phenomenon. In the first part of the chapter, we shall present our results regarding the adsorption and diffusion of molecules on the Au(111) surface using the standard DFT approach. Later on we shall compare these results with the ones obtained with the vdW-DF method which accounts for dispersion interaction. We shall present the results in the same order as in the previous chapter starting from the melamine and then moving towards the more complicated PTCDA, PTCDI, NTCDI and NTCDA molecules.

4.1 Melamine on the Au(111) Surface

In this section, theoretical results concerning the interaction between melamine networks and the Au(111) surface will be presented. In particular, we shall investigate if there are any preferential adsorption sites for melamine on the Au(111) surface and determine the diffusion barriers for a melamine molecule to move across the surface. There is also a more important issue for which we shall not be able to give a satisfactory solution, related to the incommensurability between the lattices of gold and of the gas-phase melamine in the network formation on the surface; only a preliminary analysis of it is presented.

M. Mura, *Self-Assembly of Flat Organic Molecules on Metal Surfaces*, Springer Theses, DOI: 10.1007/978-3-642-30325-8_4,

Table 4.1 Energetics of a melamine molecule on the Au(111) surface: comparison between slabs containing two and three layers of gold

	Two layers	Three layers
E_{ads}(eV)	−0.18	−0.19
E_{def} (eV)	0.02	0.02
E_{int}(eV)	−0.20	−0.21
Height (Å)	3.5	3.5

The height of the molecule above the surface in its planar configuration is shown in the last row

Since the standard DFT methods underestimate the effect of the non-local dispersion interaction which could be essential in considering the molecules on the surface, we performed two different kinds of calculations: the standard DFT method, and classical molecular dynamics. Classical molecular dynamics has been considered in order to include the dispersion energies as well.

4.1.1 Adsorption of Melamine on the Au(111) Surface

We start by describing our DFT calculations of a single melamine molecule on the Au(111) surface. The adsorption energies are reported in Table 4.1, while the top and side views of the adsorption geometry are presented in Fig. 4.1. Most of the calculations were performed with a slab of two layers of the gold atoms since we did not find any considerable difference with the calculations involving 3 layers of gold, see Table 4.1. We have used the same setup as in [1].

We tried different initial positions of the melamine molecule with respect to the gold lattice, by placing the molecule at 3.5 Å above the surface (the optimum distance found in our DFT calculations). In all cases, after the relaxation, the hydrogens of the amino group are shifted towards the gold surface. For this reason, the melamine is not planar and it is also not parallel to the surface, see Fig. 4.1. However, in all cases considered, the adsorption energy is small and changes very little (by no more than 0.05 eV, i.e. up to 30 % of the adsorption energy) with the lateral position of the molecule; some preference may be given to a position in which the N atom of an amino group of melamine is directly above a Au atom of the surface. However, the energy gain here, corresponding to about 5 % of the adsorption energy, is within the precision of DFT. Hence, according to our DFT calculations, there is no preferential adsorption site for the melamine on the Au(111) surface and the melamine should be able to move quite easily across the Au(111) surface at room temperature (RT).

A similar calculation was done for the melamine dimer, and we found the same behavior with a small adsorption energy which changes insignificantly with the dimer lateral position. One can see that the dimer (considered as one body) adsorption energy E_{ads} given in the second column in Table 4.2, is almost twice the adsorption energy of a single melamine, reported earlier in Table 4.1. This means that the adsorption energy of a cluster of melamine molecules on the gold surface is expected

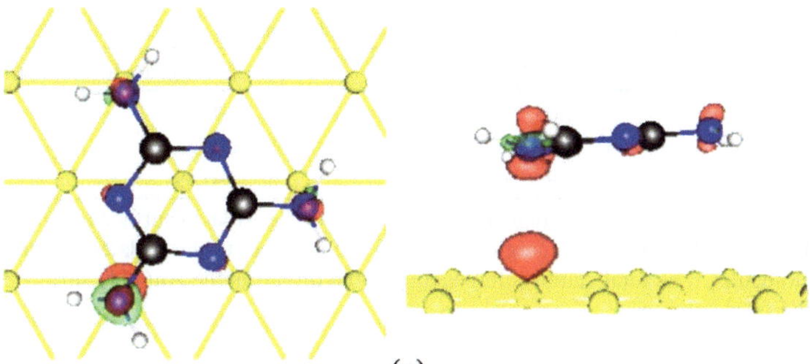

(a)

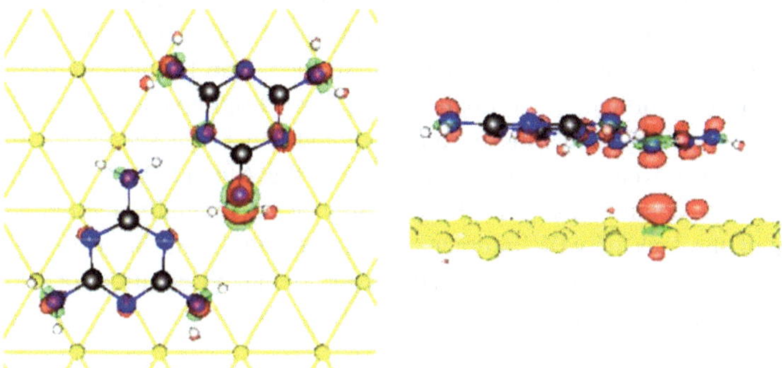

(b)

Fig. 4.1 The *top* (the *left column*) and *side* (the *right*) views of the electron density difference plots of a melamine molecule **a** and a melamine dimer **b** on the Au(111) surface corresponding to ±0.01 electrons/Å^3. The *green* surfaces correspond to the regions of positive electron density difference (excess) and the *red* areas correspond to the regions of negative electrons density difference (depletion). Some rather weak interaction of one of the melamine N atoms with the Au atom underneath it is visible in both cases resulting in slightly bent geometries

to be proportional to the number of molecules in the cluster. We also see from the last column in Table 4.2 that the interaction of the dimer with the gold surface does not affect the stabilisation of the double hydrogen bond in the dimer, and the hydrogen bond energy remains the same. Indeed, the total stabilisation energy of the whole system given in the last column of Table 4.2, is the sum of the binding energies of the two melamine molecules to each other and to the gold surface. By subtracting the two stabilisation energies given in the upper row of the table, we get −0.47 eV for the dimer stabilisation energy on the surface, which is close to the value calculated for the dimer in the gas phase, −0.48 eV, presented in Table 3.1. These calculations clearly show that the gold surface does not affect the hydrogen bonds stabilising the melamine dimer. In other words, one may say that, according to our DFT calculations, the surface provides a rather weak potential for melamine molecules in the direction perpendicular to the surface, and which, at least for practical purposes, is also completely homogeneous laterally.

Table 4.2 Energetics of the melamine dimer on the Au(111) surface

	Dimer-Au(111)	Melamine–melamine-Au(111)
E_{ads}(eV)	−0.30	−0.77
E_{def}(eV)	0.09	0.12
E_{int}(eV)	−0.39	−0.89

In the second column the energies calculated with respect to the isolated surface and the dimer are presented. In the third column the energies are given with respect to the isolated surface and two melamine molecules. The total stabilisation energy of the dimer on the Au(111) surface is the top number in the third column

It is well known that the PBE density functional we used in our calculations, as described above, does not take account of the vdW interaction between molecules, as well as between the molecules and the surface. The former contribution is not expected to be significant as compared to the dimer binding energy already provided by DFT (−0.44 eV), and thus it can be thought to be of secondary importance. However, a very weak binding of melamine to the Au(111) surface (−0.15 eV per molecule) was obtained using DFT and it may be substantially and underestimation because of the lack of vdW interaction. Another important point is that the small adsorption energy is inconsistent with the experimental observations of the desorption temperature that normally is around 100–330 °C [2–5].

4.1.2 Commensurability of the Melamine Network and the Au(111) Surface

We will now investigate the problem of commensurability of the melamine network and the Au(111) surface. This is a general and complicated problem related to different lattices of the substrate and the gas-phase monolayer of molecules adsorbed on it. For instance, the melamine hexagonal geometry presented in Sect. 3.1.1.3 which we obtained in the gas-phase calculations does not match that of the Au(111) surface itself. Although both structures are hexagonal, the lengths of the lattice vectors are different: 10.66 Å in the gas-phase melamine network and 3.00 Å of the gold surface (our theoretical value). The ratio between these lattice vectors is not a rational number. On the other hand, if we consider a supercell, i.e. a cell of such size that the network is commensurate with the substrate, the lattice vectors of the supercell must be some linear combinations of the basic lattice vectors of the gold surface because of the existence of the bulk underneath the gold surface (i.e. the 2D space group of the surface network must be a subgroup of the 2D space group of the gold surface). However, it is not clear how one can find the "best" periodicity for the network within the periodic boundary conditions (PBC), since, in practice, when performing calculations of the network with the substrate explicitly included, the lattice vectors of the supercell must be predefined as some linear combinations of the substrate lattice vectors. Taking different linear combinations may result in a different ratio between the numbers of the substrate atoms and adsorbed molecules in the cell. This makes the comparison between these calculations nontrivial. Note that reduction of the space symmetry due to a network formation can in fact be substantial in some cases. For instance, as it was observed in Ref. [6], in some cases, including the melamine on the Au(111) surface, supercell lattice vectors as seen in STM images may contain tens of the substrate lattice vectors.

The presence of the surface could influence in some way the structure of the melamine network. To investigate this, we have adopted the following method. We have considered a supercell of gold which is commensurate in a certain direction with the lattice of one of the melamine chains based on a two-molecule cell. By expanding the gold supercell two-fold, the same chain with a double cell (four molecules) can then be accommodated. In the gas phase both distances between any pair of neighbouring molecules are the same, whereas when the chain is allowed to relax on the surface by the DFT method, this will no longer be true: the chain with two molecules in the cell would have two different intermolecular distances, while the chain with four molecules will have four. Therefore, by measuring the variation of distances between molecules in the chain relaxed on the surface we can learn about the possible effects of it on the chain periodicity. Similarly, if we consider angles formed by centres of three consecutive molecules along the zig–zag shape of the chain in the gas phase, these will all be the same. However, if we measure these angles along the chain relaxed on the surface, there will be some variations: for the chain containing two molecules in the cell, there will be two independent angles, while for the chain with four molecules in the cell-four. Therefore, the variation of the angles along the chain can also be used to characterise the possible effects of the surface on the chain periodicity. In the same spirit, one could also look at the interatomic distances of each melamine molecule in more detail, however, this has not been attempted here.

The results are collected in Table 4.3. Examples of periodic systems used in these calculations for all three chains with four molecules in the unit cell are shown in Fig. 4.2. Consider the first Chain1, its gas-phase lattice vector is 10.66 Å, see Table 3.3. When placed on the surface, the chain lattice vector is defined by the chosen gold supercell and cannot be changed. It is possible to choose such gold cells where the cell vectors are close to the melamine chains lattice vectors differing by no more that 0.8 %. (We considered several orientations of the chain with respect to the gold; in each case an appropriate cell of the gold was chosen.) From our gas-phase calculations we know that the distance between adjacent molecules is 6.41 Å, while the angle along the zig–zag shape of the chain is 115.1 °. When the chain is relaxed on the surface, we find that the distances between molecules differ by ± 0.1 Å, while the angles vary by 4.6 °, which are 1.6 and 4 % of the gas-phase values, respectively. Corresponding analysis undertaken for Chain2 shows similar deviations from the gas phase values, see Table 4.3. Deviations for Chain3 are much larger, especially in the angle. Note that the deformation energies of these chains (calculated with respect to the gas-phase chains) are nevertheless very small, smaller than the DFT adsorption energies (0.01–0.09 eV/molecule and up to 0.23 eV/molecule for one of the orientations of Chain3, where the deformation is large).

Table 4.3 Variations (in % of the gas-phase value) of the lattice vectors, inter-molecular distances and angles along the zig–zag shape of the three chains relaxed with DFT on the Au(111) surface

	Chain1		Chain2		Chain3[a]	
	Gas-phase	Variation	Gas-phase	Variation	Gas-phase	Variation
Lattice vector	10.70 Å	0.8 %	12.26 Å	2.2 %	7.76 Å	4.5 %
M–M distance	6.41 Å	1.6 %	6.43 Å	3 %	6.57 Å	3 %
M–M–M angle	115.1 °	4 %	170.2 °	1.1 %	172.5 °	18 %

Some of the periodic systems are shown in Fig. 4.2. (In the case of Chain3 only the system with four molecules in the cell was considered.)

In the calculations described so far the structures of the chains were constrained by the lattice vectors of the gold supercells chosen to accommodate the chains. Unfortunately, when trying to repeat these calculations for melamine 2D periodic structures on gold, we were unable to find appropriate gold supercells of small enough size without distorting the melamine 2D networks considerably. Therefore, in order to perform an analysis similar to that done for the chains, we instead considered finite clusters of six molecules, shown in Fig. 4.3, corresponding to the hexagonal and close-packed networks of Sect. 3.1.1.3. The clusters were relaxed using our DFT method (and, therefore, with the vdW interaction not included). The inter-molecular distances and angles formed by three consecutive molecules were found to be 6.6 ± 0.3 Å and $178 \pm 2°$, respectively, i.e. the deviations are still within 3 % of the gas-phase values for the same clusters.

The important conclusion one can draw from these results is that distance and angular variations, even when small, are not completely negligible, so that the deviation from periodicity during the network formation can be accumulated, resulting in a distorted periodic (quasiperiodic) arrangement of molecules. It is worth noting, in this respect, that the orientation of chains and, therefore, monolayers, with respect to the substrate, is more sensitive to the variation in the angle than in distance. This might be quite important for networks whose character is different from that of the substrates (e.g. close-packed on top of hexagonal); and, indeed, in the case of Chain3 we find a very significant angle variation of up to 31 ° which should help in matching the close-packed arrangement (based on this chain) with the underlying gold lattice.

4.1.3 Corrugation of the Surface Potential

To investigate the corrugation of the surface potential for the melamine molecule and hence calculate the corresponding diffusion barriers, we placed the molecule at a distance of 3.5 Å above the gold surface (the DFT optimum height) and translated it in steps of 0.1 Å along several directions across the surface ($[\bar{1}10]$, $[10\bar{1}]$ and $[11\bar{2}]$), see Fig. 4.4; during each relaxation, performed using our DFT method, the lateral position of one melamine atom was fixed, although the atom was allowed to move vertically. All other atoms of the systems, including the upper layer atoms of the

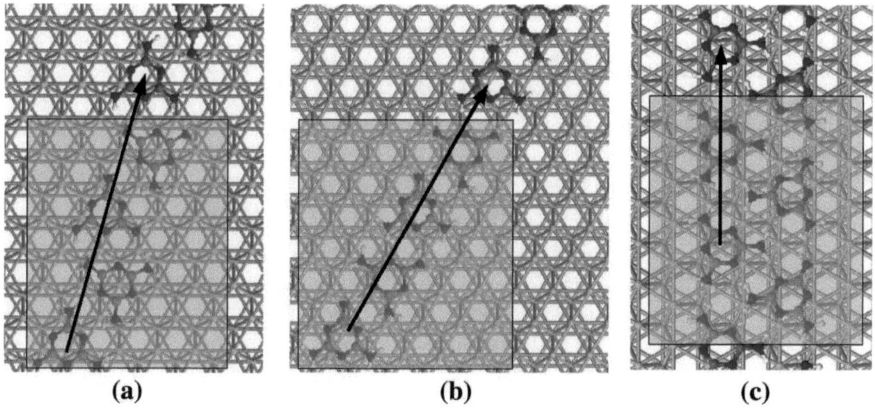

Fig. 4.2 Periodic systems containing Chain1 (**a**), Chain2 (**b**) and Chain3 (**c**) on the Au(111) surface. The unit cells containing four molecules are explicitly indicated in all cases

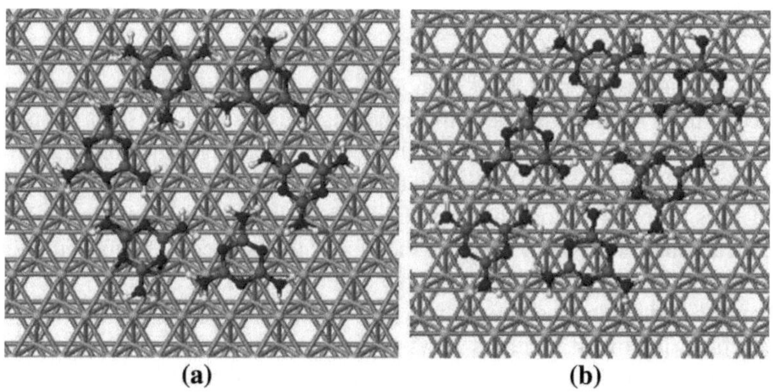

Fig. 4.3 Hexagonal (**a**) and close-packed (**b**) clusters of six molecules relaxed on the Au(111) surface

surface, were allowed to relax. We also considered the rotation of the melamine molecule around its centre of mass, see Fig. 4.4.

These DFT calculations show that the lateral change of the binding energy is very small, never exceeding 0.05 eV, i.e. 30 % of the DFT binding energy, see Fig. 4.5. These calculations confirm the high mobility of the melamine on the Au(111) surface and that it is not possible to find a preferential adsorption site for it. This means that DFT is also incapable of explaining preferential directions of the melamine island formation observed in experiment [6–8].

As DFT lacks vdW interactions between the molecules and the surface, we have also considered the role that vdW interaction may play in the corrugation of the surface potential. Inclusion of the vdW interaction via the force field [9] as described in Sect. 2.10, resulted in a similar but smaller corrugation of 0.02 eV when the height

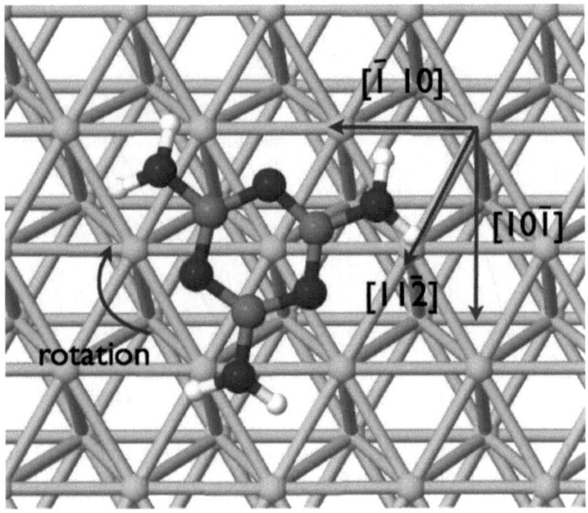

Fig. 4.4 Schematic description of different directions used to move the melamine across the Au(111) surface and the considered direction of rotation of the melamine around its centre of mass

of the molecule was fixed at 3.5 Å (the DFT obtained separation) and the molecule remained planar. When we repeated these calculations at the closer separation of 2.8 Å which corresponds to the minimum of the force field energy, the corrugation of the surface potential increased only slightly to 0.06 eV, nearly the same value as obtained by DFT; this time corresponding to around 5 % of the binding energy. This kind of classical calculation was also done for the dimer; the barrier was approximately two times larger than for a single molecule but still remained very small.

When the molecule was considered in a non-planar geometry (i.e. not parallel to the surface, its lowest H atoms 2.3 Å above the surface and the highest N and H atoms 3.4 Å above), then the situation changed drastically: the barriers for diffusion increased ten-fold reaching almost 0.6 eV and the adsorption energies jumped to values ranging between −2.3 to −1.7 eV for a non-planar H-bonded dimer. As vdW calculations using molecular dynamics are quite approximate, we cannot say at this point whether the planar or bent geometry is the most stable. It is clear, therefore, that more precise calculations which include properly the vdW interaction between molecules and the surface are required to explain the preferential adsorption of planar molecules on flat metal surfaces.

4.2 PTCDA/PTCDI and NTCDA/NTCDI on the Au(111) Surface

PTCDA, PTCDI, NTCDA and NTCDI molecules have been largely studied because they have interesting chemical and physical properties, such as fluorescence and semi-conductance. The following step in our analysis is to understand how these molecules interact with the surface. As done for the melamine molecules, we placed the single molecules in random positions with respect to the surface. Then we used

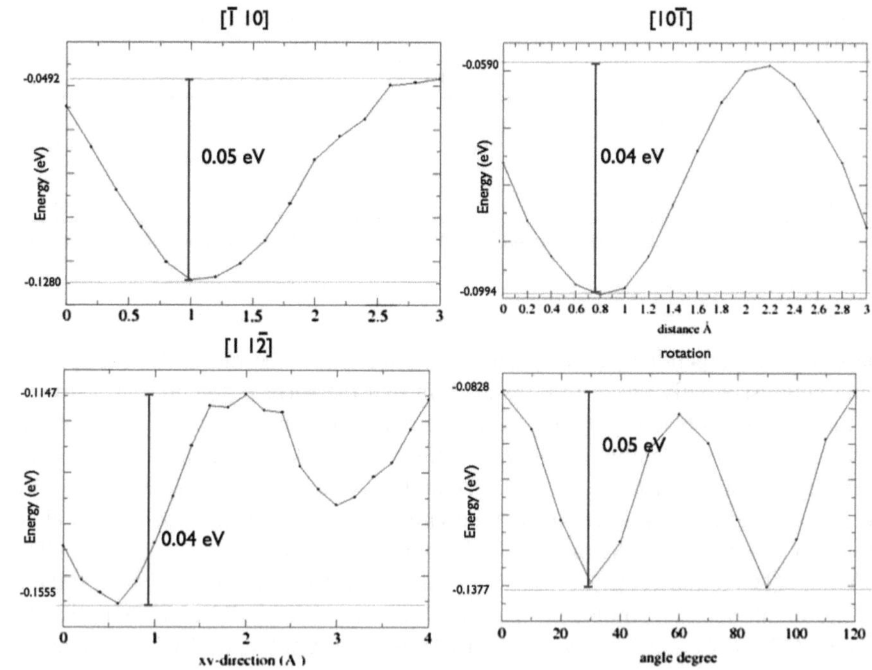

Fig. 4.5 Diffusion of the melamine molecule on the Au(111) surface calculated with our DFT method along three surface directions and the rotation of a melamine around its centre of mass

our DFT method to calculate the adsorption energy of each molecule on the Au(111) surface. We investigated in more detail the role of the vdW forces on the interaction energy between molecule and surface using standard classical molecular dynamics. Also, a part of this section is dedicated to the study of the diffusion of the molecules on the Au(111) surface.

4.2.1 Adsorption of PTCDA, PTCDI, NTCDA and NTCDA on the Au(111) Surface

PTCDA, PTCDI, NTCDA, and NTCDI molecules have the same core but they differ in the number of atoms and terminations at the short sides. We start describing our DFT calculations of single molecules on the Au(111) surface. We have used the same setup as in the melamine case and in [10–12] with a slab of two layers of gold atoms.

Next, we placed single molecules melamine, NTCDA and PTCDA at random positions in flat geometries on the surface at a distance of around 3.0–3.5 Å from it. After the geometry relaxation we find, in all cases, that the molecules almost remain in the original lateral positions. The only considerable displacement is observed along

Table 4.4 Energetics of PTCDA, PTCDI, NTCDA and NTCDI molecules on the Au(111) surface (in eV)

	PTCDA	PTCDI	NTCDA	NTCDI
E_{ads}	−0.17	−0.13	−0.10	−0.15
E_{def}	0.06	0.06	0.11	0.10
E_{int}	−0.23	−0.19	−0.21	−0.25
Height	3.5	3.6	3.6	3.6

The height of the molecules above the surface in their planar configurations is shown in the last row, (in Å)

the direction perpendicular to the surface along which the molecule is displaced, remaining essentially flat. These results hint that the surface potential, as provided by the PBE, must be laterally flat. After the relaxation, the molecules lie flat on the surface at the considerable distance of about 3.5–3.6 Å, see Table 4.4 (the experimentally measured distance for PTCDA on the Au(111) surface is 0.23 Å smaller [13]). The binding energies, shown in the same table, were found to be very small, of the order of −0.1/−0.2 eV; moreover, the binding energies of rather large PTCDA, PTCDI, NTCDI and NTCDA molecules have been found to be nearly the same as for the small melamine molecule. The BSSE corrections, in all cases, are noticeable and are 0.23, 0.24 and 0.35 eV for the melamine, NTCDA and PTCDA molecules, respectively.

The electron density difference plots, see Fig. 4.6, show a weak molecule-surface interaction. We show only the PTCDA, PTCDI and NTCDA electron density difference plots as we expect that we would have a similar behaviour for NTCDI.

Also in this case we find in our calculations very small adsorption energies which do not depend on the molecular dimension. NTCDA, NTCDI, PTCDA and PTCDI are much larger molecules compared with the melamine, however, most of the energies obtained are smaller than the adsorption energy of the melamine molecule on the Au(111) surface, see Tables 4.1 and 4.4. The small values of the adsorption energies of the molecules on the Au(111) surface, obtained using standard DFT, are due to the lack of the vdW interactions (dispersion forces).

In the next section we shall discuss diffusion of the molecules on the surface using the same method as for the melamine molecule.

4.2.2 Corrugation of the Surface Potential

To investigate the corrugation of the surface potential for the perylene molecules and hence calculate the corresponding diffusion barriers, we placed the molecules 3.5 Å above the gold surface (the DFT optimum height) and translated it in step of 0.1 Å along several directions across the surface ($[\bar{1}10]$, ($[10\bar{1}]$) and ($[11\bar{2}]$), as it was done for the melamine molecule.

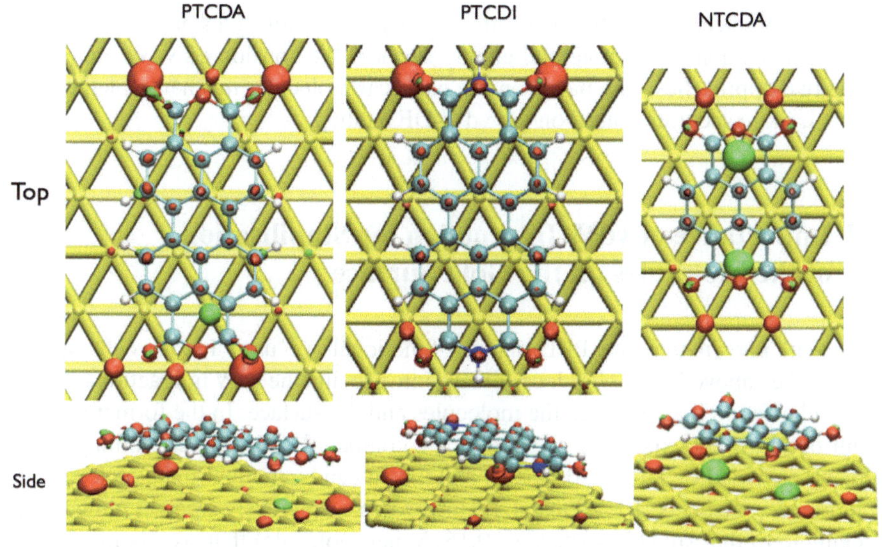

Fig. 4.6 *Top* and *side* views of the electron density difference of PTCDA, PTCDI and NTCDA molecules on the Au(111) surface corresponding respectively to ±0.01 electrons/Å^3 for PTCDA, ±0.004 electrons/Å^3 for the PTCDI and NTCDA molecules. The *green* surfaces correspond to the regions of positive electron density difference (excess) and the *red* areas correspond to the regions of negative electrons density difference (depletion)

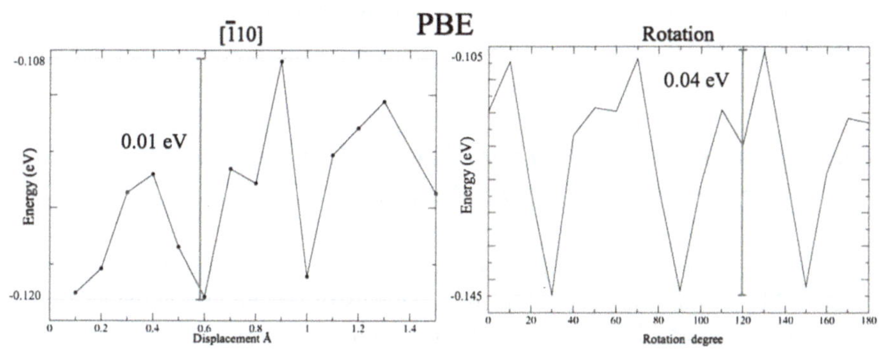

Fig. 4.7 The relative energy of a PTCDA molecule calculated using PBE method as a function of the molecule displacement along the surface direction (as indicated) and when rotating the molecule around its centre of mass

We performed these DFT calculations for PTCDA and PTCDI, we find that the lateral change of the binding energy is very small for these molecules, with the corrugation of the potential energy surface of such large molecules never exceeding 0.04 eV, see Fig. 4.7.

We note that the vdW interaction may play a role in the corrugation of the surface potential. In the next section we shall investigate in more detail the role played by the the vdW interaction in the adsorption energy of the molecules on the surface and in the corrugation of the potential energy surface. We considered vdW interaction using two approaches: one based on the classical molecular mechanic method and another, more accurate, based on the vdW-DF method.

4.3 Importance of vdW Interaction for Stabilisation of the Molecules on the Gold Surface

It is well known that the PBE density functional we used in our calculations, as described above (Sect. 2.7), does not take into account the vdW interaction between molecules, as well as between the molecules and the surface. In the former case the contribution is not expected to be significant compared to the dimer binding energy already provided by DFT (−0.44 eV), and thus it can be thought to be of secondary importance, see Sects. 3.1.1.2 and 4.1. However, a suspiciously weak binding of e.g. melamine to the Au(111) surface (−0.15 eV per molecule) that we obtained using DFT may be substantially underestimated because of the lack of the vdW interaction. It is possible, at present, to take account of the missing vdW interaction [14–16] and this is what we shall discuss here.

In order to evaluate the contribution of vdW dispersion forces, we considered two different ways to proceed: one is an approximate classical method based on the Sci-Fi code [17] and another one is an ab initio, more accurate method based on the vdW-DF method.

4.3.1 Approximate Method with Sci-Fi: Analysis of Adsorption Energy and Corrugation Potential

Initially we have attempted a simpler approach in which we employed a force field [9] developed by fitting results of the quantum chemistry calculations at the MP2 level of various groups of all DNA base molecules with gold clusters. Note that the force field includes all interactions between the bases and the gold surface, including vdW. In these calculations, performed using the Sci-Fi code [17] we did not relax either the surface, nor the melamine molecule and only the interaction energy between the two was accounted for. Hence, the total energy of the melamine-surface system represents directly the adsorption energy. The surface was represented as a finite $20 \times 20 \times 4$ cluster of gold atoms which, we have checked, was large enough to neglect the effect of its boundary on the energetics of a single melamine molecule positioned somewhere around its centre. The minimum adsorption energy for the melamine was found to be at the molecule-surface separation of 2.9 Å and amounted to almost −1.1 eV, which represents quite a considerable value of the binding energy. This energy minimum is very different compared with the results shown in Sect. 4.1.1 where, after

Table 4.5 Interaction of single molecules with the Au(111) surface studied using the classical potentials [9]

Molecule	Binding energy, eV	Height above the surface, Å
Melamine	−1.08	2.9
NTCDA	−1.31	3.1
NTCDI	−1.58	3.2
PTCDA	−2.03	3.0
PTCDI	−2.14	3.2

the geometry relaxation with our DFT method, for a melamine on the Au(111) surface, the molecule-surface separation was 3.5 Å, with an adsorption energy of 0.15 eV. A similar approach has been applied to PTCDA, NTCDA, NTCDI and PTCDI molecules as well. In all cases the energy minimum of the molecules on the Au(111) surface is found to be at different molecule-surface separations compared with the values obtained with DFT, see Table 4.4, and in all cases the adsorption energies are increased considerably, see Tables 4.5 and 4.4.

We also used this method to evaluate the diffusion barrier of the molecules on the surface. In order to perform these calculations we considered the same surface directions as in the DFT calculations, see Fig. 4.4.

Most importantly and similarly to the DFT calculations, these values did show no evident dependence on the lateral position of the molecule, at least if the molecule remains planar. Thus, we conclude that the addition of the vdW interaction, even at this approximate level, does result in significant binding energies. However, our main conclusion, related to the corrugation of the surface potential, remains largely unaffected by this, see Fig. 4.8.

In the next section we shall verify these results with a more accurate ab initio vdW-DF method.

4.3.2 vdW-DF Method (Quantum Espresso and Siesta) Applied to Molecules on the Au(111) Surface

Bearing in mind the noticeable cost of the calculations due to the large system sizes considered, and the fact that, in this study, we would mainly like to draw qualitative conclusions on the importance of the dispersion interaction for the molecule-surface systems, all calculations of molecular adsorption on the gold surface were performed using a slab with two layers of gold only, in which the bottom layer atoms were fixed in the bulk geometry, while the upper layer atoms were all free to relax. We carefully checked that using three layers of gold (with either one or two layers allowed to relax) changes the binding energies of a melamine molecule on gold only marginally, as described above in Sect. 4.1.1. The lateral dimensions of the unit cells were in all cases chosen such that the interaction between images of adsorbed molecules is

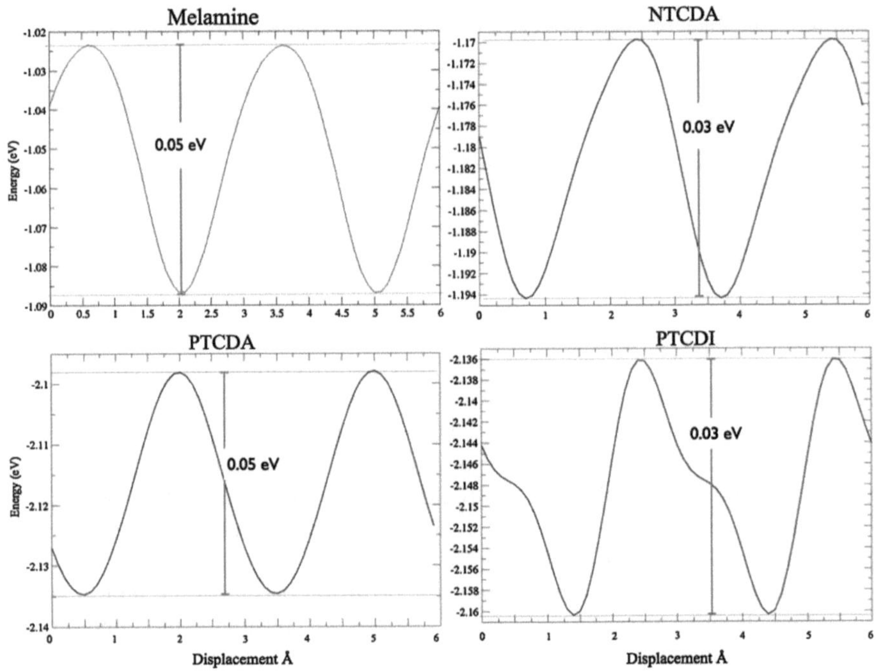

Fig. 4.8 The relative energy of a molecule (melamine, NTCDA, PTCDA and PTCDI) calculated using our molecular mechanics method (Sci-Fi) as a function of the molecule displacement along the surface $[\bar{1}01]$ directions on the Au(111) surface

negligible: in the case of the adsorption of a single melamine on gold, a 4×3 gold cell (24 atoms in each layer) was used, while for the melamine dimer, NTCDA, and PTCDA adsorption we used a 4×5 gold cell (40 atoms in each layer). The largest system we considered, a PTCDA on the two-layer gold surface, contained 118 atoms.

The electronic charge density difference for the PTCDA molecule on the gold surface was calculated by subtracting the individual electron densities of the molecule and the surface, both calculated in the geometry of the combined system, from the density of the latter. This is shown for the PBE calculation in Fig. 4.9a which clearly demonstrates that there is no density accumulation between the surface and the molecule, although some redistribution of the density within each subsystem is visible. This finding is also supported by the Mulliken population analysis. These conclusions agree with similar calculations of the melamine and adenine (A) molecules on the same surface [1, 11], and are also confirmed by the projected electronic density of states (PDOS) (see discussion below). Thus, according to the PBE based calculations, the binding mechanism for the flat molecules considered here on this metal surface, is a weak physisorption.

As anticipated, this picture changes dramatically if the dispersion interaction is accounted for. For comparison, two types of calculations have been performed:

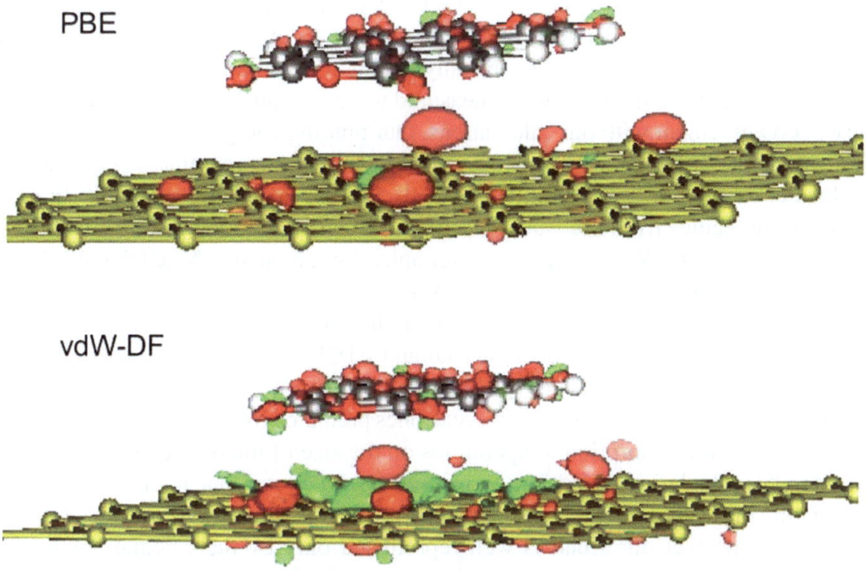

Fig. 4.9 Electron density difference for the PTCDA molecule on the Au(111) surface calculated using PBE (*top*) and vdW-DF (*bottom*) methods. *Red* and *green contours* represent depletion and excess (at ±0.004 electrons/Å^3) of electron density, respectively. Only the *upper layer* of the gold surface is shown. The aim is to give a complete characterisation of the supramolecular networks, taking into account the balance between the molecule–molecule and molecule–substrate interactions

Table 4.6 Interaction of single molecules (of N atoms) with the Au(111) surface studied using the PBE and vdW-DF methods

Molecule	N	Binding energy, eV				Height above the surface, Å			
		PBE	Non-scf	scf	Classical	PBE	Non-scf	scf	Classical
M	15	−0.25	−0.71	−0.88	−1.08	3.5	3.5	3.1	2.9
NTCDA	24	−0.10	−1.06	−1.31	−1.31	3.6	3.6	3.5	3.1
PTCDA	38	−0.17	−1.70	−1.88	−2.03	3.6	3.5	3.3	3.0
NTCDI	26	−0.15	–	−1.20	−1.58	3.6	–	3.3	3.2
PTCDI	40	−0.13	–	−1.94	−2.14	3.5	–	3.2	3.2

In the latter case we show results of both electronically self-consistent (marked "scf") and non-self-consistent (marked "non-scf") calculations. Results of the simulations using the classical potentials [9] are also given under "Classical" for comparison

non-self-consistent (the PBE electron densities and geometries were used) and self-consistent (fully vdW-DF calculations). The binding energies of the same molecules on the gold surface, calculated self-consistently in the same way as above but using the vdW-DF method and shown in Table 4.6, become quite significant. Moreover, the binding energy starts to depend on the size of the molecule in accord with the intuitive picture that the vdW interaction is *additive*: the binding energies of the three molecules relate to each other as 1.4:1.0:0.7 which is close to the corresponding relationship 1.6:1.0:0.6 between the numbers of their atoms. As far as we know, experimental adsorption energies are only available for the NTCDA and PTCDA molecules on gold [18], which were measured to be 1.5 and 2.0 eV, respectively, in a very good agreement with our calculations. Our binding energies are also supported by an indirect experimental information on the desorption temperatures of melamine, naphthalene tetracarboxylic diimide (NTCDI), and perylene tetracarboxylic diimide (PTCDI) molecules from the Au(111) surface which were found to be 80, 200, and 330 °C, respectively (W. Xu Privater communication). Note that NTCDI and PTCDI molecules are very similar to the NTCDA and PTCDA molecules considered here, so that desorption temperatures for the latter should be expected to be of the same order of magnitude. The results of NTCDI and PTCDI are shown in Table 4.6. Two observations can be made: (i) the vdW-DF binding energies are more consistent with desorption temperatures than the very weak ones predicted by PBE; (ii) there is a clear dependence of the desorption temperatures on the size of the molecules, and, again the vdW-DF calculations yields the correct results. Of course, one has to be careful in associating the desorption temperatures directly with the adsorption energies of *single* molecules; however, the trend is well reproduced only by the calculations which account for the dispersion interaction. Note that the BSSE corrections are significant (0.24, 0.28, and 0.42 eV for the melamine, NTCDA, and PTCDA, respectively) and are of the same order of magnitude as the PBE calculations. Due to the cost of the calculations and, because here we are only interested in qualitative conclusions, bigger basis set calculations, which would reduce the BSSE, were not tried.

Interestingly, the results for the binding energies obtained in non-self-consistent calculations, as it is clear from Table 4.6, are all systematically lower than those obtained in fully self-consistent calculations. The underestimation of binding varies between 0.17–0.18 eV for melamine and PTCDA molecules and 0.25 eV for NTCDA molecules which are of the same order as the total binding energies obtained in our PBE based calculations. These results emphasise the importance of full self-consistency in vdW-DF based calculations.

To investigate whether the surface potential is indeed flat, we performed extensive diffusion calculations for melamine and PTCDA molecules on the surface, in which a single carbon atom in their benzene-like core was moved in small steps along several directions on the surface with subsequent relaxation of all other atoms of the molecule and the upper layer atoms of the surface; the vertical component of the chosen atom was also allowed to relax. The total displacements for the molecules considered across the surface were up to 3 Å. These calculations gave a variation of the binding energy of less than 0.03 eV as a function of the molecules lateral position confirming the fact that the surface potential is indeed extremely flat.

We performed these calculations using the vdW-DF method by translating the molecules in 0.05 Å steps along several directions across the surface ($[\bar{1}10]$ and $[10\bar{1}]$) as indicated in Fig. 4.10. The results of these calculations are shown in Fig. 4.11 for the melamine and Fig. 4.12 for the PTCDA molecules, respectively. In some of the

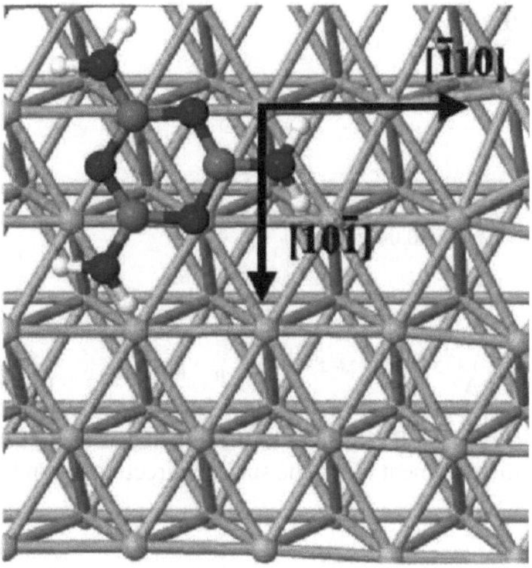

Fig. 4.10 The starting geometry of the melamine molecule on the Au(111) surface. The *red arrows* indicate different directions considered to move the molecule across the surface

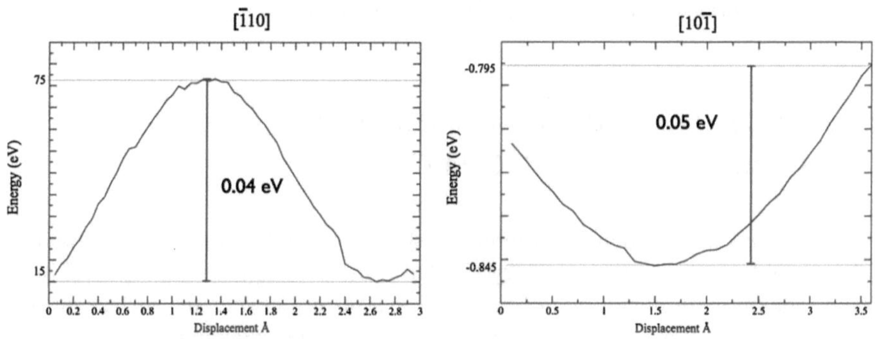

Fig. 4.11 The relative total energy of a melamine molecule as a function of its position on the surface calculated using the vdW-DF functionals

calculations the steps taken were rather big, so that the curves connecting the points are only used to guide the eye.

These results confirm an intuitive view that, at least for the gold surface studied here, the vdW interaction simply provides stronger binding energies; in fact, this is the main binding mechanism. Otherwise, the electron density of the molecule-gold systems is not significantly modified by the dispersion interaction, and there is no significant change in the electronic DOS as well, compared with the PBE based calculations, as discussed in the next section.

The geometries of the relaxed systems did not change significantly: the molecules still lie flat above the surface, but at a somehow closer distance than in our PBE calculations (see Table 4.6). In fact, in the case of PTCDA, our calculated distance

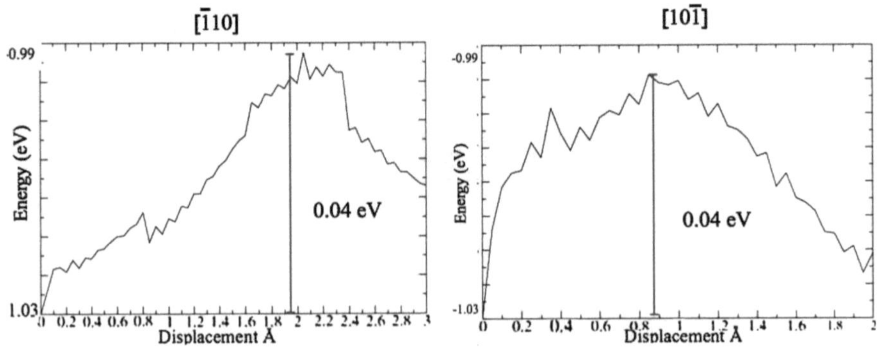

Fig. 4.12 The relative energy of a PTCDA molecule calculated using the vdW-DF methods as a function of the molecule displacement along the surface directions as indicated

of 3.3 Å is now much closer to the experimental one of 3.27 Å [13]. Overall, atomic displacements within the molecules are not larger than 0.01 Å. Note that, since the revPBE exchange tends to give longer bond lengths [19], we may expect that our molecule-to-surface distances are slightly overestimated.

4.3.3 Effect of vdW-DF Functional in the Electronic Charge Density Difference: PTCDA Case

The calculated electron density differences, shown in Fig. 4.9b for the PTCDA molecule on gold as an example, are very similar to those obtained with the PBE functional. The previous conclusion of zero charge transfer between the surface and the molecules firmly stays in place in the case of the vdW-DF calculations as well. This result is also confirmed with the PDOS analysis, shown for the PTCDA molecule on gold in Fig. 4.13. The DOS projected on the PTCDA atoms in the combined system is extremely close to the DOS of an isolated molecule (indicating on a small hybridization of the molecular and metal states), and the LUMO (the lowest unoccupied molecular orbital) state of the molecule lies above the Fermi level of the metal, and thus remains unoccupied, suggesting no charge transfer. A similar result was obtained also in our PBE calculations (not shown). Interestingly, as seen from Table 4.6, the binding energies obtained in the vdW-DF based calculations for all three cases considered here are very close to those obtained in classical calculations using the force field developed in [9]. However, the force field based calculations predict slightly smaller distances between the molecules and the gold surface.

In order to assess the effect of electronic self-consistency on the electron density when performing vdW-DF calculations, we considered the configuration of the PTCDA molecule on the gold surface relaxed using the vdW-DF method, and calculated the electron density of this system using the PBE functional. If a non-self-

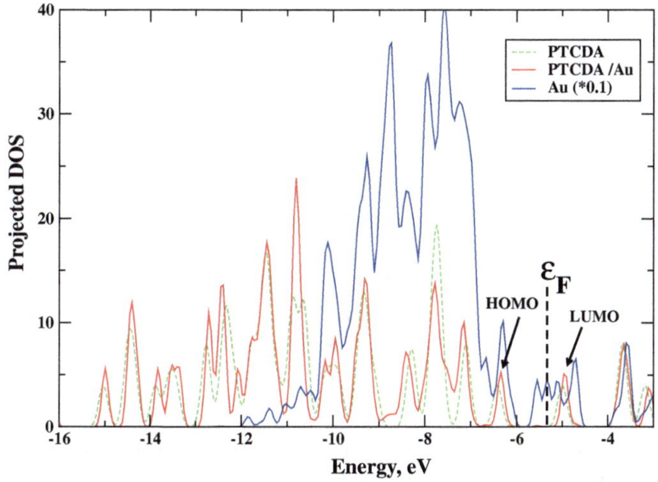

Fig. 4.13 Electronic projected density of states (PDOS) for the PTCDA molecule on the Au(111) surface calculated with the vdW-DF method using a single Γ point and the Gaussian smearing with 0.2 eV dispersion. The projections are shown on atoms of the molecule (PTCDA/Au, *red*) and of the surface (Au, *blue*); the contribution of the surface was reduced 10 times for convenience. The part of the DOS for energies lower than −16 eV is not shown. The Fermi energy of the metal surface is indicated. For comparison, the total DOS of an isolated PTCDA molecule is also shown by the (*green*) *dashed line* with HOMO and LUMO orbitals indicated explicitly

consistent calculation was performed, this PBE electron density would have been used. The change of the electron density due to self-consistency for this system is shown in Fig. 4.14. Note that the same contour levels are used here as in Fig. 4.9 which shows electron density difference due to adsorption. The main effect for the gold surface is found to be in some density redistribution around the Au atoms, which is not related to the molecule adsorption. Note that, only green contours corresponding to density accumulation around gold atoms are visible whereas density depletion regions (red) are more localised and hence not visible; the total charge density change within the surface amounts to zero. Further, there is no visible change in the electron density between the molecule and the surface. However, one can notice some accumulation of the density around all the atoms of the molecule with subsequent depletion of it in the regions between them, including the regions inside the rings. The absolute values of the density change are not big (between −0.009 and 0.074 electrons/Å^3); note, however, these are of the same order of magnitude as due to the adsorption itself.

Finally, we have considered a melamine dimer on the gold using the vdW-DF method. It was shown previously in [11] that a PBE based calculation does not predict any change in the hydrogen bonding between the two molecules due to their (rather weak) interaction with the gold surface, and the overall binding energy of two melamine molecules to the gold is basically equal to the sum of the isolated dimer binding energy and twice the single melamine adsorption energy. Similar results

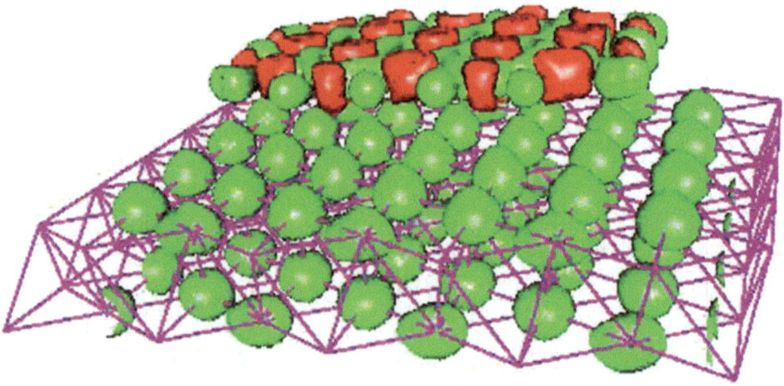

Fig. 4.14 The change of the electronic density of the PTCDA molecule on the gold surface, calculated using the vdW-DF functional, due to self-consistency. *Red* and *green contours* represent negative and positive changes (at ±0.004 electrons/Å^3) of the electron density, respectively

have been obtained here with the vdW-DF method as well; however, the binding to the surface was found to be much stronger. Indeed, the total binding energy of the dimer to the gold surface, calculated with respect to the isolated surface and two separate melamine molecules, has been found to be equal to −2.08 eV. If we add the melamine-melamine dimer binding energy of −0.42 eV, Table 3.27, and twice the adsorption energy for a single melamine molecule of −0.88 eV, Table 4.6, we would get a very similar energy of −2.18 eV. One can also clearly see the well developed "kebab" structure between the two melamine molecules and the absence of the density in the space between the molecules and the gold in the electron density difference plot for this system in Fig. 4.15. In fact, the kebab structure is practically unchanged as compared with that for the isolated melamine dimer (not shown). This again demonstrates an additivity of the interactions in this system. We observe a very small polarization of the density of the surface, however, no charge transfer to/from the surface is present.

4.4 Conclusions

We have studied the adsorption of a selection of flat organic molecules of variable sizes on the Au(111) surface using two density functionals: PBE, which lacks the dispersion interaction, and the vdW-DF, which is completely parameter-free, but accounts for this interaction in an approximate way. All our calculations are fully self-consistent and the vdW interaction was also accounted for in calculating atomic forces. A recently developed [20] efficient implementation of the vdW-DF method in the SIESTA code [21] was used.

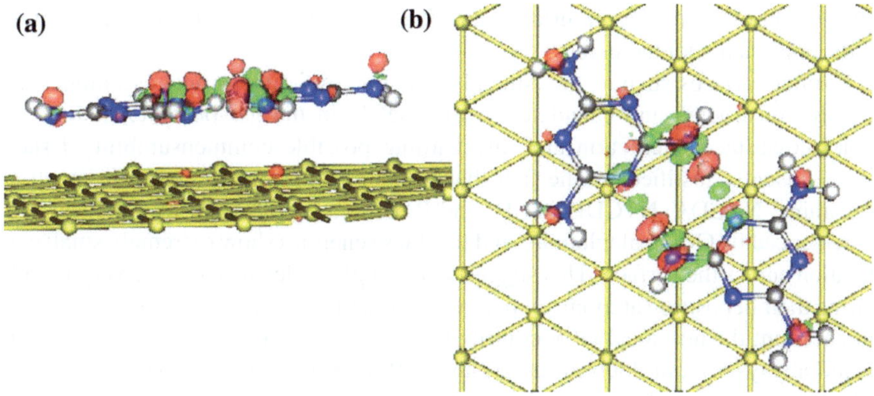

Fig. 4.15 *Side* (**a**) and *top* (**b**) views of the electronic density difference plot for the melamine dimer on the Au(111) surface obtained by subtracting the densities of isolated molecules and of the surface, in the geometries of the combined system, from the density of the latter. The *red* and *green contours* correspond to depletion and excess (at ± 0.004 electrons/Å ^{3}) of the electron density, respectively. Only the *upper layer* of the gold surface is shown

We find that the vdW interaction affects very little the geometries of the molecules on the gold surface, as compared to the PBE based calculations. The only significant difference is that the molecules lie slightly closer to the surface due to vdW interactions.

However, as anticipated, binding (adsorption) energies of these flat molecules to the Au(111) surface are effected significantly by the dispersion interaction. While the PBE functional predicts very weak binding which does not depend on the lateral sizes of the molecules, the dispersion interaction, as implemented in the vdW-DF method, drastically changes this picture: the binding energies increase many times with the dispersion interaction providing most of the binding. We also find that the binding energy depends linearly on the size of these planar molecules, i.e. larger molecules experience larger vdW interaction to the gold surface with respect to smaller ones. This is in agreement with the available experimental data and intuitive understanding that the vdW interaction is additive by nature. In addition, we have also considered a melamine-melamine dimer adsorbed on the gold surface. We find that the vdW interaction does not affect the hydrogen bonding within the dimer. However, as expected, the adsorption energies increase dramatically.

We have also investigated the role of electronic self-consistency in performing vdW-DF calculations. We found that the change in the electronic density due to self-consistency is of the same order of magnitude as its change for a particular system due to adsorption. Moreover, we find that the binding energies obtained in non-self-consistent calculations are all systematically lower than the ones obtained in fully self-consistent calculations, with the discrepancies reaching 20 % in two of the cases. These results emphasise the importance of performing fully self-consistent calculations of molecules on surfaces, and hence provide an fundamental justification of the approach adopted here.

For flat surfaces such as the Au(111), possible two-dimensional assemblies are usually constructed and calculated (within DFT) in the gas-phase excluding the molecule-surface interaction and disregarding possible commensurability issues. This is usually justified by the fact that the interaction of the molecules such as Melamine, NTCDA, NTCDI, PTCDI and PTCDA are very weak with this surface, as evidenced by GGA calculations, and that these energies show extremely small corrugation across the surface. This suggests that single molecules must be very mobile during their deposition at room temperature. Since these DFT calculations do not take into account the non local vdW contribution to the correlation energy, these results would always remain doubtful unless the vdW contribution is accounted for. PBE calculations suggest that molecules on the gold surface can freely move across the surface, however, their evaporation temperatures, due to weak binding, are expected to be very low in strong disagreement with available observations. The vdW interaction corrects this issue and provides significant binding; the flatness of the surface potential still remains in place, i.e. the molecules can be considered as "floating" in a 2D pool: they cannot easily escape from the surface, but can freely move around. This is in agreement with the STM observations that it is impossible to image small flat molecules on the gold surface at room temperature at small coverages [1, 22, 23]. Our results also imply that the gas-phase modeling, at least in the first approximation, should be adequate for this surface.

References

1. Kelly REA, Xu W, Lukas M, Otero R, Mura M, Lee YJ, Lægsgaard E, Stensgaard I, Kantorovich LN, Besenbacher F (2008) An investigation into the interactions between self-assembled adenine molecules and the au(111) surface. Small 4:1494
2. Saywell A, Magnano G, Satterley CJ, Pedigão LMA, Champness PH, Beton NR, O'Shea JN (2008) J Phys Chem C 112(20):7706–7709
3. Perdigão LMA, Champness NR, Beton PH (2006) Chem Commun 5:538
4. Perdigão LMA, Fontes GN, Rogers BL, Oxtoby NS, Goretzki G, Champness NR, Beton PH (2007) Phys Rev B 76:245402
5. Perdigão LMA, Perkins EW, Ma J, Staniec PA, Rogers BL, Champness NR, Beton PH (2006) J Phys Chem B 110:12539
6. Staniec PA, Perdigão LMA, Rogers BL, Champness NR, Beton PH (2007) J Phys Chem C 111:886
7. Zhang H, Xie Z, Long L, Zhong H, Zhao W, Mao B, Xu X, Zhao W (2008) J Phys Chem C 112:4209
8. Xu W, Dong M, Gersen H, Rauls E, Vázquez-Campos S, Crego-Calama M, Reinhoudt DN, Stensgaard I, Laegsgaard E, Linderoth TR, Besenbacher F (2007) Small 3:854
9. Piana S, Bilic A (2006) J Phys Chem B 110:23467
10. Silly F, Shaw AQ, Castell MR, Briggs GAD, Mura M, Martsinovich N, Kantorovich LN (2008) J Phys Chem C 112:11476
11. Mura M, Martsinovich N, Kantorovich LN (2008) Nanotechnology 19:465704

12. Mura M, Gulans A, Thonhauser T, Kantorovich L (2010) Role of van der waals interaction in forming molecule-metal junctions: flat organic molecules on the au(111) surface. Phys Chem Chem Phys 12(18):4759–4767
13. Henze SKM, Bauer O, Lee T-LT-L, Sokolowski M, Tautz FS (2008) Surf. Sci. 601:1566–1573
14. Dion M, Rydberg H, Schroeder E, Langreth DC, Lundqvist BI (2004) Phys Rev Lett 92:246401
15. Thonhauser T, Cooper VR, Li S, Puzder A, Hyldgaard P, Langreth DC (2007) Phys Rev B 76:125112
16. Cooper VR, Thonhauser T, Puzder A, Schroeder E, Lundqvist BI, Langreth DC (2008) J Am Chem Soc 130:1304
17. Kantorovich LN, Trevethan T, Polesel-Maris J, Foster A. Self consistent image force interaction + virtual AFM machine. http://www.cmmp.ucl.ac.uk/~lev/codes/SciFi/manual_3_51/index.html
18. Ziroff J, Gold P, Bendounan A, Forster F, Reinert F (2009) Surf Sci 603:354–358
19. Perdew JP, Burke K, Ernzerhof M (1998) Phys Rev Lett 80:891
20. Gulans A, Puska MJ, Nienminen RN (2009) Phys Rev B 79:201105
21. Soler JM, Artacho E, Gale JD, Garcia A, Junquera J, Ordejon P, Sanchez-Portal D (2002) Phys Condens Matter 14:2745
22. Otero R, Lukas M, Kelly REA, Xu W, Lægsgaard E, Stensgaard I, Kantorovich LN, Besenbacher F (2008) Elementary structural motifs in a random network of cytosine adsorbed on a gold(111) surface. Science 319:312–315
23. Kelly REA, Lukas M, Kantorovich LN, Otero R, Xu W, Mura M, Laesgaard E, Stensgaard I, Besenbacher F (2008) Understanding disorder of the dna base cytosine on the au(111) surface. J Chem Phys 129:184707

Chapter 5
Influence of Dynamics of Melamine with Au ad-Atom on the Au(111) Surface on Self Assembled Structures: Bright Spots

In the previous chapters we analysed the self-assembly process due mainly to the combination of molecule-molecule and molecule-surface interactions. It is clear from the previous chapter that for flat organic molecules deposited on the Au(111) surface the dispersion interaction plays a crucial role in the binding of the molecules to the surface, but this interaction does not affect the corrugation of the surface potential which remainsvery small. This suggests that molecules during their deposition at room temperature cannot easily escape from the surface, but they can freely move around the surface.

In our previous discussion we assumed that the surface remains intact during the self-assembly process and it is merely represented by planar terraces. In this chapter we address a somewhat different situation, in which one has to assume, as it will be clear later on, that freely diffusing surface ad-atoms also have to be accounted for during the deposition process of molecules leading to their assembly. Specifically, we considered a situation observed experimentally when additional bright features may appear within the gaps between the molecules. In Fig. 5.1 the STM images of the hexagonal domains of melamine molecules deposited on the Au(111) surface are shown. The experiment has been conducted by Dr. Fabien Silly from Oxford University (unpublished). He found that under particular experimental conditions one can observe the hexagonal network of melamine with or without bright spots between the molecules. In order to explain the results of these experiments one has to analyse the deposition process from a dynamical point of view. The bright spots also have been observed for the same network and on the same surface from Zhang et al. [1]. The presence of the bright spots has been also observed in the self-assembly of benzenethiol on the Au(111) surface [2] where the Au adatoms present on the Au(111) surface are presumably involved in the self-assembly mechanism. The spots are observed after the co-deposition of iron atoms and 1,3,5-tricarboxylic benzoic acid (trimesic acid, TMA) on Cu(100) [3] and after the deposition of PTCDA on the Ag(110) surface [4].

The first question that one has to answer is why in some cases the bright spots appear. One possibility could be that these are due to electronic effects of the tip

M. Mura, *Self-Assembly of Flat Organic Molecules on Metal Surfaces*, Springer Theses, DOI: 10.1007/978-3-642-30325-8_5,

Fig. 5.1 STM images of the hexagonal domain of melamine on the Au(111) surface in which either the bright (**a**) or dark (**b**) spots appear in the network between the triangular features corresponding to melamine molecules. Note also occasional dark spots in image (**a**) and bright spots in (**b**)

during the scanning of the sample. However, this explanation does not stand up to scrutiny. Indeed, the spots are periodically distributed in a large portion of the sample. When the spots are present the network is slightly distorted and this could suggest the presence of some impurities in the centre of each hexagon of the network. Dr. Fabien Silly, who performed the experiment, noticed that the phenomenon is apparently correlated with the coverage of the molecules on the surface: he observed the bright spots at a low coverage of 0.1 monolayer and in small islands while for a high coverage of 1.0 monolayer black spots, instead of the bright ones, appear. Experimentally, the low coverage was achieved by using small deposition rate of the melamine molecules, which high deposition rate of the molecules was used in obtaining high coverage networks.

The presence of the bright spots in the hexagonal network of the melamine has been analysed in the paper by Zhang et al. [1]. After the preparation of the sample they observed either black or bright spots in the centres of the hexagons in domains of a melamine hexagonal network. They associated black spots with empty units and the bright dots with "filled" units. Combining DFT calculations and geometrical considerations [5, 6] they concluded that the presence of bright or black spots is due to the presence or absence of melamine molecules that stand perpendicularly to the surface (i.e. up-right). In their analysis Zhang et al. [1] tried to justify the presence of the bright spots but they did not discuss any correlation between the presence (or absence) of the bright spots and the melamine coverage. Considering the correlation between the coverage and the presence of the bright spots on the network it is clear that the hypothesis of Zhang et al. [1] can not be the right explanation because their hypothesis would imply that the bright spots should be observed for high coverages only, which contradicts the experiments by Dr. F. Silly.

Our assumption is that gold atoms are trapped in the network and that the deposition rate and the coverage play important roles in the formation of the network itself. At a low coverage of 0.1 monolayer there is a small number of melamine molecules and a certain amount of Au ad-atoms; as the melamine molecules diffuse on the substrate, there is a high probability for the Au ad-atoms to move around and be bound with melamine molecules and hence form a hexagonal network with "trapped" Au atoms in its pores; at high coverage, however, there is a huge number of melamine molecules which are free to diffuse on the substrate, and there is a higher chance for the melamine molecules to meet and build the network, while there is lower probability for the gold to be trapped.

In order to understand the physical processes involved, we have to answer a few questions about what happens on the surface and to perform both relaxation and diffusion calculations.

The next section is dedicated to the consideration diffusion of the Au ad-atom on the Au(111) surface. We then investigate the possibility for the Au atoms to stabilise the network and we consider the interaction of molecules or clusters with ad-atoms on the surface. Particular care should be taken in order to understand the network formation when the presence of gold ad-atoms is accounted for.

5.1 Diffusion of Au Atoms on the Au(111) Surface and Detachment from a Step Edge

The Au(111) surface is not perfectly clean (presents a certain amount of impurities) and flat: step and islands serve as sources (and drains) of the gold atoms. As a result, there will always be some equilibrium concentration of diffusing gold ad-atoms on terraces at the given temperature which is determined from the detailed balance of the rates at the step edges and islands. Kinetic Monte Carlo (KMC) simulations [7] can be useful in obtaining this equilibrium, given a certain distribution of islands and steps. KMC is an essential method in understanding the dynamics of melamine hexagonal lattice formation with the gold atoms present on the surface. This kind of calculation, although going beyond the scope of this thesis (due to the lack of time), requires knowledge of not only the diffusion barriers of single gold atoms on the terrace, but also their barriers for the attachment and detachment to/from the steps and islands. Since the diffusion barriers for a single melamine molecule were already considered in Chap. 4, we shall concentrate here on the diffusion of individual gold atoms.

Using the same method as for studying the melamine diffusion, we obtained the barrier for the diffusion of the Au atom on the terrace to be about 0.1 eV (compare with Refs. [8, 9]) along the $[\bar{1}10]$ and $[1\bar{1}0]$ directions on the surface. This is still

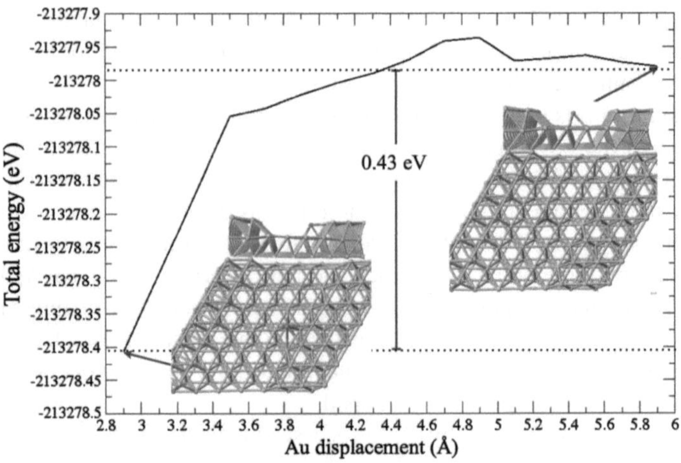

Fig. 5.2 The total energy of the Au ad-atom (shown in *pink*) as a function of its displacement from the step edge

a small barrier, although it is twice as large as the 0.05 eV barrier for the melamine diffusion.

We have also investigated the Au atom detachment from the step edge using the DFT method. The initial configuration is the one where the gold ad-atom is bonded to the step edge and the bond length is 3.0 Å. Then we displaced the Au atom in the $[\bar{1}10]$ direction across the surface away from the step edge, see Fig. 5.2. In the final configuration the Au atom is detached from the step edge, being at the distance of 5.9 Å from the step edge.

The calculated energy that one has to overcome in order to detach one Au atom from the step-edge is found to be 0.43 eV. The calculated barriers can be used to estimate the equilibrium concentration of freely moving gold atoms on the surface.

5.2 Gas Phase Calculations: First Attempts

In order to estimate the stabilisation of the network in the presence of the gold atoms we initially performed some gas phase calculations. Using the same procedure as in Sect. 3.1, we relaxed the hexagonal domain of the melamine with an added gold atom in the centre of each hexagon of the network. We chose two possible positions of the Au ad-atom: the first one with the gold atom below the network by around 0.5 Å and the other one by 1.5 Å. These two configurations have been chosen considering the relative height of the Au ad-atom with respect to the Au(111) slab and the melamine network. The hexagonal network was, in the first configuration, only slightly deformed after the relaxation, while in the second case it was deformed more significantly, see Fig. 5.3a. In both cases the Au atoms relaxed in the centre of

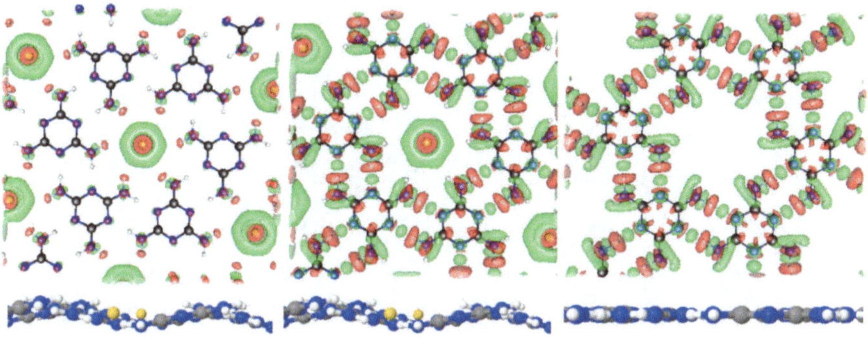

Fig. 5.3 Electron density difference, corresponding to ±0.01 electrons/Å^3, of the hexagonal melamine network with ((**a**) and (**b**)) and without (**c**) gold atoms. The electron density difference is shown: **a** between the total density and the densities of the individual melamine and gold networks; **b** between the total density and the individual densities of all molecules and gold atoms; **c** the same as (**b**), but without the gold atoms. The *green* surfaces correspond to the regions of positive electron density difference (excess) and the *red* areas—to the regions of negative electron density difference (depletion). The side view of each network is shown at the *bottom* of the corresponding density difference plot

each hexagon and the network modified appropriately in order to keep the gold in the centre of each hexagon. The presence of the gold atoms in the network induces also a rearrangement of the electronic density difference, see Fig. 5.3.

In both cases the presence of the gold atom increases the stability of the network.

The next point to consider is the effect of the surface in this formation of the network. In order to have a good understanding of what happens at the surface, we performed our calculations in a systematic way. Our starting point is related to the interaction between the single melamine molecule on the surface and an individual gold ad-atom.

5.3 Interaction of a Melamine with an Au ad-Atom on the Au(111) Surface

In this section we shall investigate the role played by the surface in the stabilisation of the system "melamine + Au ad-atom". In order to evaluate the interaction between the melamine and a Au ad-atom on the Au(111) surface, we first relaxed two configurations using the DFT method in which one melamine and one Au ad-atom were placed on the surface. In one geometry the gold is close to the molecule and in the second geometry the gold is far away from it, see Fig. 5.4. The role of these two calculations is to estimate the stabilisation of the "molecule-Au ad-atom" complex or block. In the first configuration (close) the relaxed distance between the gold and

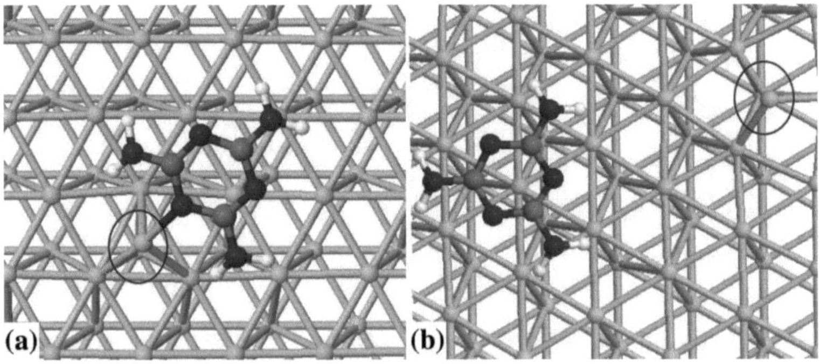

Fig. 5.4 The geometries of a melamine and a Au ad-atom on the Au(111) surface: **a** close and **b** far away. The ad-atom is surrounded by a *red circle* for clarity

the melamine is $d_{\min} = 2.23$Å, measured by the gold ad-atom and the closest nitrogen atom of the melamine. In the second case the ad-atom was far enough to avoid any interaction between them, $\approx$7.7Å. Energetically the close configuration is more favourable by 0.9 eV with respect to the far configuration.

This calculation demonstrates that the block "melamine + Au ad-atom" is very stable: once the complex is formed, it is very difficult to break, a barrier of around 0.9 eV needs to be overcome.

Next we performed some diffusion calculations in order to investigate the bond formation and possible orientations of the melamine. We started from the configuration in which the melamine and the Au ad-atom are at a distance of 6.5 Å and then we translated the melamine towards the Au atom using small steps of 0.1 Å. During each relaxation, performed using our DFT method, the lateral position of the furthest carbon atom of the melamine was fixed, although it was allowed to move vertically.

We started from the particular configuration shown in Fig. 5.5, then, exploiting the periodic boundary conditions, we translated the molecule in the opposite direction across the surface towards the Au atom. These two diffusion processes correspond to the melamine moving along $[\bar{1}10]$ and [110] surface directions.

After relaxation, the distance between the melamine molecule and the Au ad-atom is 2.24 Å for orientation 1 and 2.43 Å for orientation 2. In both cases the molecule is bonded to the Au ad-atom but orientation 1 results in a more stable configuration by 0.2 eV. As seen in Fig. 5.5, when the melamine is moved towards the Au ad-atom, they form a single entity and afterwards move together as one block "melamine + Au ad-atom", keeping the distance of 2.25 Å. Next we performed a translation of the whole block "melamine + Au ad-atom" on the Au(111) surface, and the corrugation energy for this system was found to be 0.2 eV. To perform the diffusion of the block "melamine + Au ad-atom" we started from the final configuration of orientation 1 and then pushed the melamine even closer to the Au ad-atom; we find that they move and relax together maintaining the distance of 2.25 Å.

We believe that orientation 1 is a stable minimum of the PES, while orientation 2 is a saddle point, because of the symmetry of this geometry and the fact that any rotation of the melamine in either direction would result in more stable geometry

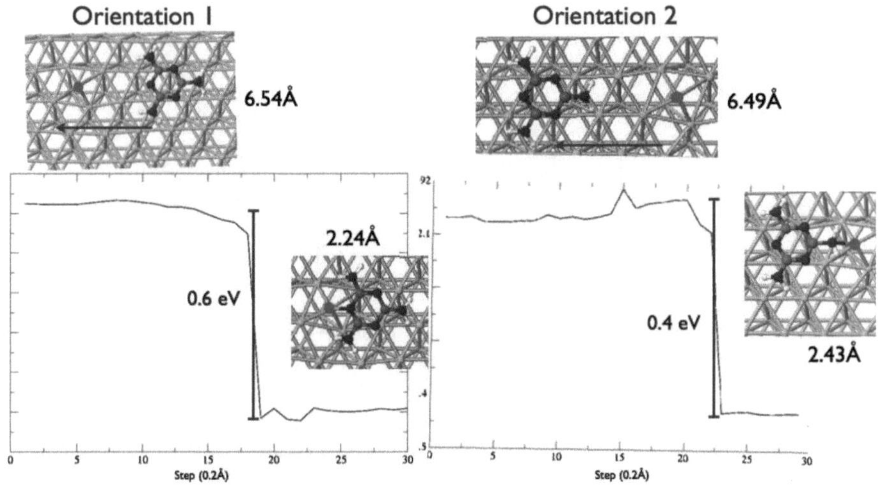

Fig. 5.5 Total energy of a melamine and a Au ad-atom (shown in *pink*) as a function of the diffusion steps for two orientations of the melamine. The starting and final geometries for the two orientations are also shown (see insets)

corresponding to orientation 1. Therefore, orientation 2 can be discarded from the following analysis as a state of the system which may only exist for a very short time.

If there is a low coverage of melamine molecules on the Au(111) surface, free Au atoms available on the surface would form complexes with the melamine molecules as in orientation 1 prior to melamine network formation. They can then easily diffuse on the surface as single entities, exposing the other two sides of the melamine for possible hydrogen bond formation with other melamine molecules. When other molecules come close, the network will start to grow via formation of hydrogen bonds with the other available sites of the melamine in the complex. Then the gold atom may move to the centre of the hexagons and get trapped inside them. In the case of high coverage of melamine molecules, the latter will form larger clusters faster, so that Au atoms would not have enough time to approach them and move inside the hexagons.

In order to describe in more detail the process of the network formation, we have to investigate the mechanism of hexagon formation starting from the elementary complex (or block) "melamine + Au ad-atom". Up to now we only know that the bond between the melamine and the ad-atom is strong. The strength of this bond results in some restrictions in the network formation. In the general case the melamine has six possible sites available to form hydrogen bonds distributed over its three triangular sides, however, in this case two of these sites on one molecular side are occupied by the Au atom. Considering the network formation when another melamine molecule is close to the block "melamine + Au ad-atom", the network would most likely start to grow employing the two free sides of the melamine. In the next section we describe diffusion calculations of the block "melamine + Gold ad-atom" plus a second melamine molecule in order to justify this assumption.

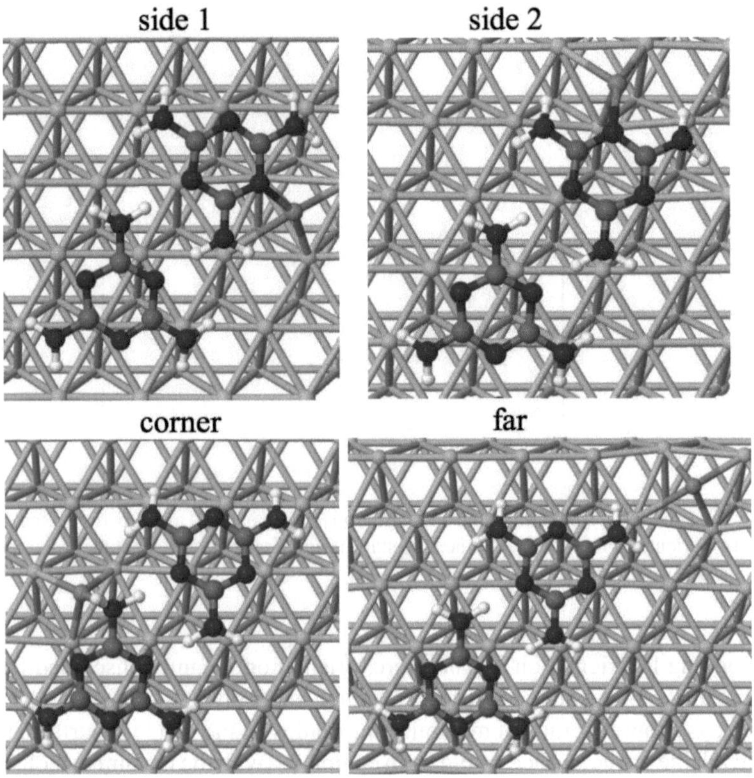

Fig. 5.6 Four possible geometries of a melamine dimer and a Au ad-atom (shown in *pink*) on the Au(111) surface: near one of the free possible sides of the melamine molecule, with the ad-atom bonded to one corner and the fourth configuration where the Au ad-atom is not bonded to the melamine

5.4 Diffusion Calculations of Melamine and "Melamine + Au ad-Atom" Block

Similarly to the previous section, we would like to assess the interaction of a melamine dimer with an Au ad-atom on the Au(111) surface. We considered four geometries with our DFT method; in three of these geometries the gold atom is in a close configuration with respect to the dimer, while in the fourth geometry the gold is far from the melamine dimer, see Fig. 5.6.

After the relaxation, we find that the most stable geometry is the one where the gold ad-atom is bonded to the side of the melamine which is not involved in the hydrogen bonds of the melamine dimer. We find no difference in energies between side 1 and side 2 configurations. These configurations are 0.75 eV more stable than the one in which the gold atom is far from the melamine dimer and cannot form the bond with the melamine. The configuration in which the Au ad-atom is bonded to the corner of one of the melamine is by 0.26 eV more stable than the far configuration; however, it is 0.49 eV less stable than any of the side configurations. Hence, we find that, when the gold ad-atom is bonded to a melamine molecule, the system becomes more stable by attaching to any of its sides near their centre, and, in addition, its presence does not affect the formation of the melamine dimer via any of the free sides of the melamine. The described calculations imply that if there is a block "melamine + Gold ad-atom" and a second molecule arrives near one of the free sides of the melamine molecule in the block, the dimer formation is likely, and may even happen spontaneously, i.e. without overcoming an energy barrier. What is not known, however, is what happens if the second melamine arrives in another direction and gets close to the side of the melamine occupied by the Au ad-atom.

To answer this question, we performed, using the DFT method, diffusion calculations where the second melamine is translated in two possible directions on the Au(111) surface towards the "melamine+Au ad-atom" block. We considered only two directions exploiting the symmetry of the melamine molecule. In each calculation the remote N atom of the second melamine molecule was moved in small steps towards the first melamine; all atomic positions were allowed to relax apart from the lateral coordinates of the chosen N atom and the bottom layer of the gold surface.

Unfortunately, when using the standard DFT method the melamine is easily de-adsorbed from the surface, because this method underestimates the interaction with the gold surface, as explained in the previous chapter. In order to overcome this problem we performed the diffusion calculations with a method that semi-empirically includes the dispersion interaction. This is done, by adding during the geometry relaxation additional interaction between the molecules and the surface due to a force field [10] to DFT forces.

We performed a diffusion calculation of the second melamine in the $[\bar{1}10]$ and $[1\bar{1}0]$ directions along the Au(111) towards the occupied side of the block "melamine + Au ad-atom". When we diffused the melamine in the $[\bar{1}10]$ direction, and the melamine gets very close to the ad-atom, it is desorbed from the surface, see Fig. 5.7, due to a considerable energy barrier in that direction.

Considering the second diffusion simulation shown in Fig. 5.8, the result is more interesting. During the first steps, the second melamine that is approaching the first one rearranges itself (positions b and c on the graph) then it has to overcome an energy barrier to arrive at a minimum, indicated as d, where the Au ad-atom is bonded to both melamine molecules. We arrive at a similar configuration by further pushing the second melamine until the next minimum, position e, is reached. Translating the second melamine even more we arrive at a maximum of energy (indicated as f) where an interchange of the gold atom between the two melamine molecules takes place. Hence, it appears that it is possible to switch the bond between the first melamine and the Au ad-atom if another melamine is approaching. The barrier for this event to happen is however rather small, only about 0.16 eV (between configurations a and f). Further diffusion of the second melamine we expect would eventually result in a formation of the melamine dimer with the gold atom being attached to one of the dimers free sides as in Fig. 5.6 (upper panels). This part of the path will require overcoming an additional energy barrier and would result in a very stable structure

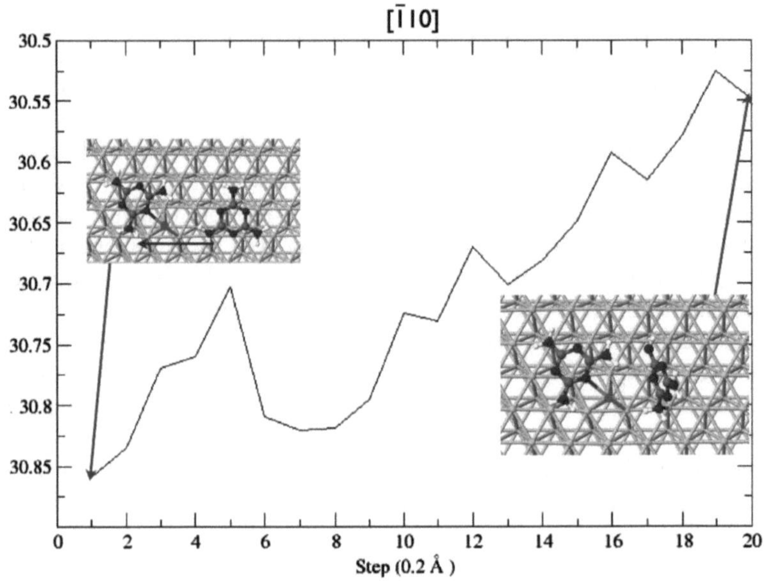

Fig. 5.7 The relative energy of a melamine molecule diffusing towards the Au ad-atom (shown in *red*) on the Au(111) surface calculated using the DFT method with a dispersion correction (see text), given as a function of the molecule displacement along the surface [$\bar{1}$10] direction. The initial and final geometries of the system are also given (inset)

with considerable energy gain. We plan to perform the corresponding Nudge Elastic Band (NEB) calculations [11–14] to evaluate the corresponding energy barriers, however, due to lack of time, these are not included in the present thesis. The important point here is, however, that the structure d (or e) with two melamine molecules binding a single gold atom is only metastable, with a barrier of only about 0.1 eV to be overcome to break it.

Summarising, with these calculations we found that if another melamine molecule approaches a "melamine + Au ad-atom" block from the side occupied by the gold atom, we observe two different behaviours that are strongly dependent on the orientation of the molecules. In either case a significant energy barrier needs to be overcome to arrive at the stable configuration in which two molecules bind to each other forming a dimer and the gold atom is attached to any of the four free sides of it. If, however, the second melamine molecule approaches the block initially from a free side of the first molecule, they bind easily forming a usual melamine dimer. This process is spontaneous as there is no energy barrier; moreover, there is a considerable energy gain, see Sect. 5.3 and also Sect. 5.5. Therefore, it is reasonable to assume that the former two processes considered above can be discarded: even if the second melamine molecule approaches the "melamine + Au ad-atom" block from the gold atom side, it still requires less energy to steer around the first melamine to

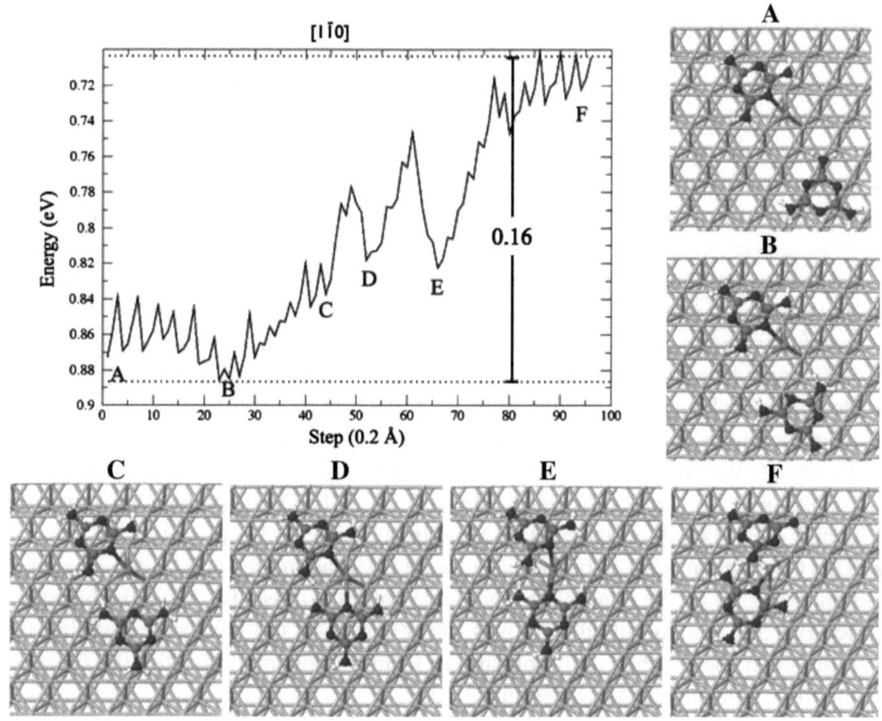

Fig. 5.8 The relative energy of a melamine molecule diffusing towards the Au ad-atom (shown in *pink*) on the Au(111) surface calculated using the DFT method with a dispersion correction (see text), given as a function of the molecule displacement along the surface direction [1$\bar{1}$0]. Geometries corresponding to some positions on the energy curve are also shown (*right*)

attach to it at one of the two free sides as this results in considerable energy gain and requires no barrier to overcome at all.

5.5 Melamine Clusters and a Au ad-Atom

So far we have characterised the interaction of the melamine molecules or a melamine dimer with the Au ad-atom on the Au(111) surface. The aim of this section is to evaluate the stabilisation energy of some melamine clusters relevant for the hexagonal melamine network formation process in the presence of Au ad-atom. To this end, we relaxed some of the melamine clusters of up to six molecules on the Au(111) surface together with the Au ad-atom, see Fig. 5.9. This sequence of clusters corresponds to a possible mechanism of a formation of a single melamine hexagon with the gold atoms inside it, see Fig. 5.10. This mechanism is based on the following points which come out from our calculations discussed above:

- gold atoms are always present on the Au(111) surface and are quite mobile on it;
- a melamine molecule diffuses freely on the surface;
- a single Au ad-atom and a melamine molecule forms a strong block when in close proximity of each other;
- that block can almost freely move around the surface as a single entity;
- if a second melamine molecule comes close to the block, it can attach to it via one of the two free sides of the first melamine; the smallest barrier for this attachment process is expected when the two melamine molecules meet with their free sides, i.e. when the bonded Au ad-atom is not on their way; if, however, the second melamine molecule comes from the gold atom side, it will steer around it to attach eventually to one of the free sides; in either case there is only a very small melamine diffusion barrier to overcome;
- the hexagonal network may therefore be formed in a step-by-step fashion: firstly, a "melamine + Au ad-atom" block is formed; then a second melamine molecule attaches to the block to one of its free sides, then the third molecule attaches to the second one, then the fourth, and so on, see Fig. 5.10. This way the whole hexagon can be constructed with the Au ad-atom being inside it, i.e. trapped.
- This mechanism is based on the assumption (to be addressed and discussed later in detail) that when the final (sixth) molecule comes to close the gap in the hexagon, see Fig. 5.9, the Au ad-atom moves to the centre of the hexagon providing the required space for the sixth molecule to form hydrogen bonds with the first and the fifth molecules as shown in Fig. 5.10.

The calculated energies of the structures shown in Fig. 5.9 are presented in Table 5.1. We found that when increasing the number of melamine molecules in the cluster, a gain in the stabilisation energy due to the additional melamine–melamine and melamine-substrate interactions is clearly seen. We also compared the energies of the clusters composed of $n + 1$ and n melamine molecules. We noticed that the difference in energy between the two consecutive clusters, $\Delta E_n = E_{n+1} - E_n$, is kept nearly constant for all structures (around 0.8–0.85 eV) except for the hexagon (note that the BSSE correction was accounted for in these calculations which were all performed within the PBE method). In the latter case there is an additional gain in the energy due to the introduction of the sixth melamine which forms hydrogen bonding with two, rather than one, melamine molecules; in addition, the translation of the Au ad-atom from the side to the centre of the hexagon position may affect the total energy of the hexagon cluster.

In order to address this latter point, we considered a melamine hexagon on the Au(111) surface with a single Au ad-atom. Following the same procedure used for the melamine and the melamine dimer, see Sects. 5.3 and 5.4, we relaxed two configurations: one with the Au ad-atom in the centre of the hexagon and another with the atom outside and away from it in order to avoid the possibility of bond formation. These calculations suggest that there is a gain in energy of approximately 0.2 eV when the Au ad-atom is in the centre of the hexagon.

Note that we could have started from the same "melamine + Au ad-atom" block and then added melamine molecules in another way, so that the Au ad-atom would

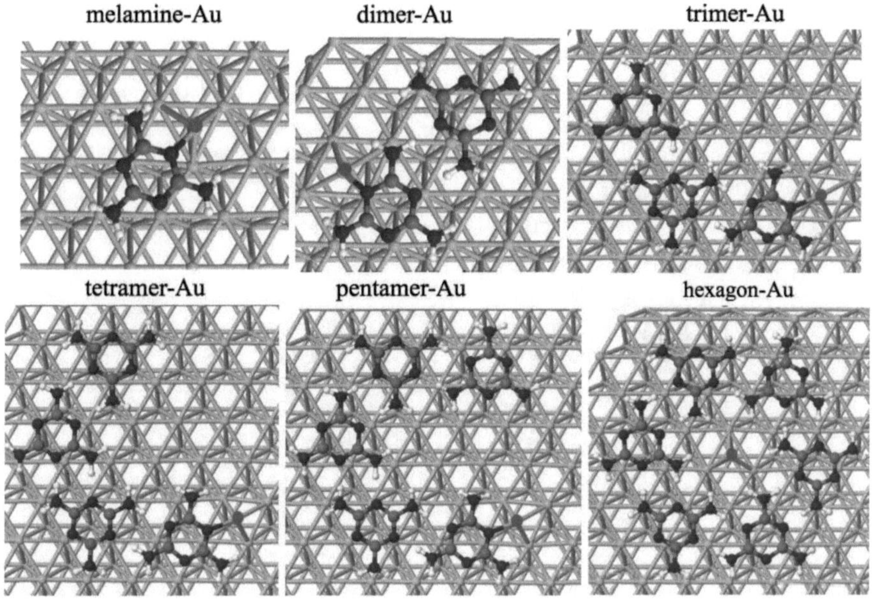

Fig. 5.9 The geometries of melamine clusters interacting with the Au ad-atom (shown in *red*) on the Au(111) surface

Table 5.1 Total energies (calculated using the PBE density functional) of different clusters of molecules with the Au ad-atom on the Au(111) surface (in eV and without the BSSE correction)

	melamine-Au	dimer-Au	trimer-Au	tetramer-Au	pentagon-Au	hexagon-Au
E_{stab}	−3.94	−4.74	−5.53	−6.38	−7.23	−8.28
ΔE_n	0	−0.80	−0.79	−0.85	−0.85	−1.05

In the second row relative energies between two consecutive clusters are given for convenience

appear *outside* the hexagon. This will happen if each time a new melamine molecule comes to a free side of a melamine in the existing cluster, and the coming molecules would close the hexagon in an appropriate way, see the right schematics in Fig. 5.11.

Summarising, we found that the hexagonal structure of the melamine molecules is more stable when the Au ad-atom is in the centre, however, we also know that the block "melamine + Au ad-atom" forms a very strong bond. On the other hand, the hexagon with the gold atom inside cannot be formed unless there is enough room for the sixth molecule to move in; for this to happen, the Au ad-atom must occupy a certain position inside the pentagon, it should detach from the first molecule to facilitate the bond formation of the latter with the sixth molecule. Depending on how the second molecule gets attached to the first, the hexagon formation may go in a different direction, as schematically shown in Fig. 5.11.

One of the three possibilities where the gold atom remains at each step outside the hexagon being formed (Model C in Fig. 5.11) is trivial and does not require any

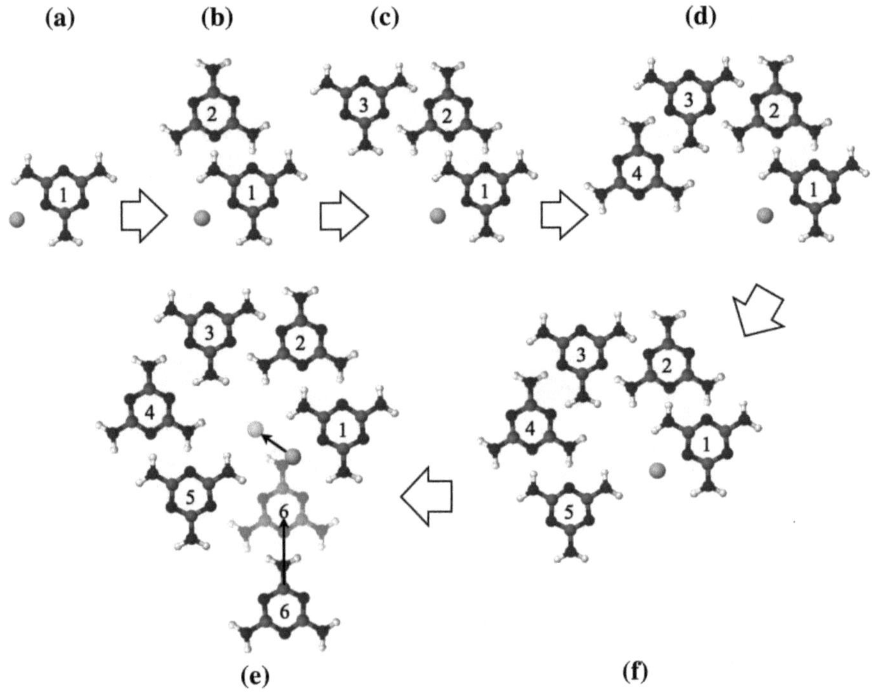

Fig. 5.10 The proposed scheme (from **a** to **f**) for the formation process of a melamine hexagon starting from **a** a "melamine + Au ad-atom" block and leading to the final **f** structure with six molecules and the gold atom in the middle

special modeling as we know from our calculations already performed and discussed above that two melamine molecules easily form a dimer on the Au(111) surface with substantial energy gain. The presence of the gold atom attached to one of them does not affect this process as long as the atom is attached to a side of the melamine not involved in the hydrogen bonding between the two melamine molecules.

In the other two scenarios (Models A and B in Fig. 5.11) the gold atom is attached to one of the end molecules of the pentagon prior to the hexagon formation and this does not allow for the sixth molecule to attach easily to close the hexagon. Note that these two cases are not identical and both require detailed modeling to assess their feasibility. Therefore, our main concern now should be about the formation of the closed hexagonal structure starting from a pentagonal cluster in which the first melamine is bonded to a Au ad-atom, and other melamine molecules form a pentagon with the first. We have to investigate whether the Au ad-atom can diffuse from its block-bonded position with the first melamine molecule outside (Model B) or to the centre (Model A) of the hexagon when the sixth molecule diffuses towards it from the outside to close the gap. In order to address this issue we performed some diffusion calculations that should elucidate the two processes.

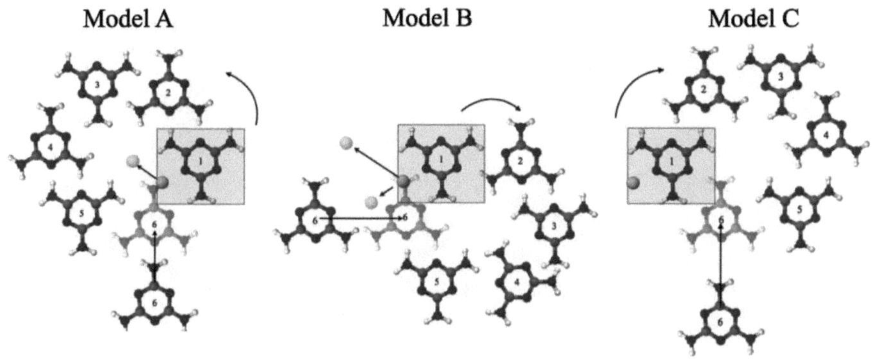

Fig. 5.11 Three main possibilities of a melamine hexagon formation from the "melamine + Au ad-atom" seed (indicated by a *square box*). The order in which the molecules attach to the seed is indicated by numbers and the *arrows*. The final positions of the sixth molecule and of the gold ad-atom are also shown as *dim*. Note that in Model B the gold ad-atom may either diffuse out of the hexagon or re-attach to the sixth molecule, as shown

5.6 The Final Stage of Formation of a Melamine Hexagon with a Au ad-Atom

The formation of the hexagonal structure with the Au ad-atom in the centre of it on the Au(111) surface is a complex problem that involves many degrees of freedom; the energy profile of this process must be complicated and it is difficult to understand the dynamics of the network formation *a priori* without considering various possible processes which might contribute to it. The essential elementary process we are addressing here is that of a formation of a closed individual melamine hexagon with a Au ad-atom attached to one of the melamine molecules; this is the system which "grew" from a "melamine + Au ad-atom" block serving as a "seed". We have seen in the previous section that three possibilities may arise (see Fig. 5.11):

- Models A and B: these are two possibilities for the Au ad-atom to appear *inside* the formed pentagon; these are due to two possible sites the gold ad-atom may occupy in the pentagon. Then there is a question of how the sixth melamine closes the gap and the complete hexagon is formed since the latter requires for the Au ad-atom to move out of the sixth molecule way. It can either move inside the hexagon to free the corresponding site of the first melamine for the sixth molecule to bind to (Model A), or it can move outside of the hexagon (Model B);
- Model C: a melamine hexagon is formed with the Au ad-atom *outside* it; this process requires a succession of simple melamine-to-melamine attachments which are spontaneous (in fact, there is a substantial energy gain) provided that the melamine molecules are sufficiently close to each other.

To shed more light on these issue, we performed calculations with our DFT method, with and without taking into account the dispersion forces. In the former case the

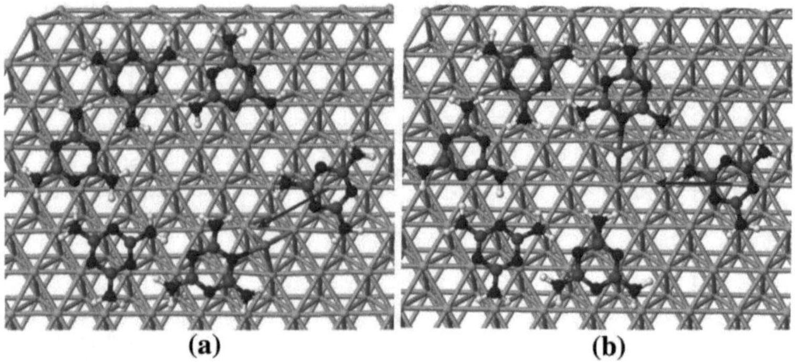

Fig. 5.12 Two possible geometries (models **b** and **a** of Fig. 5.11 respectively) of melamine pentagons with the attached Au ad-atom (shown as *pink*) and the sixth melamine molecule translated towards the pentagons in the direction indicated

dispersion interaction was accounted for semi-empirically, by simply adding to the SIESTA energy an additional interaction between the molecules and the gold surface taken from the force field [10]. The forces on the atoms were affected correspondingly. Initially we considered two clusters composed of six melamine molecules, see Fig. 5.12, and the Au ad-atom bonded in two possible ways to one of the edge pentagon melamine molecules corresponding to models A and B in Fig. 5.11. The sixth melamine molecule is considered to diffuse towards the pentagon in a direction to close the gap.

Consider model B first, see Fig. 5.12 (left). The calculations which were done with the DFT method without the dispersion correction give a weak interaction with the surface and, as a result of this, the melamine molecule easily de-adsorbs from the surface. We know from the previous chapter that in order to improve the results related to the weak adsorption energy, we have to take into account the dispersion interaction. All our calculations discussed below in this section were perfomed using the dispersion correction.

We performed diffusion calculations starting from geometry A in Fig. 5.13 and then translated the melamine molecule towards the Au ad-atom (bonded to the first melamine molecule of the cluster as shown) along the direction shown by the arrow in Fig. 5.12a. During this simulation, the most distant nitrogen atom of the sixth molecule was allowed to relax only in the vertical direction perpendicular to the surface. The total energy of the system (without the BSSE correction, which is expected to be similar along the whole diffusion path) is presented as a function of the displacement of the sixth melamine molecule in Fig. 5.13. The most stable configuration indicated with C in Fig. 5.13 corresponds to the geometry in which the Au ad-atom is bonded to two melamine molecules at the same time. We have come across this geometry in the previous Sect. 5.5). Further displacements of the sixth molecule, after overcoming a barrier of around 0.2 eV, leads to its re-orientation, and consequently the Au ad-atom gets re-attached from the fifth molecule to the sixth.

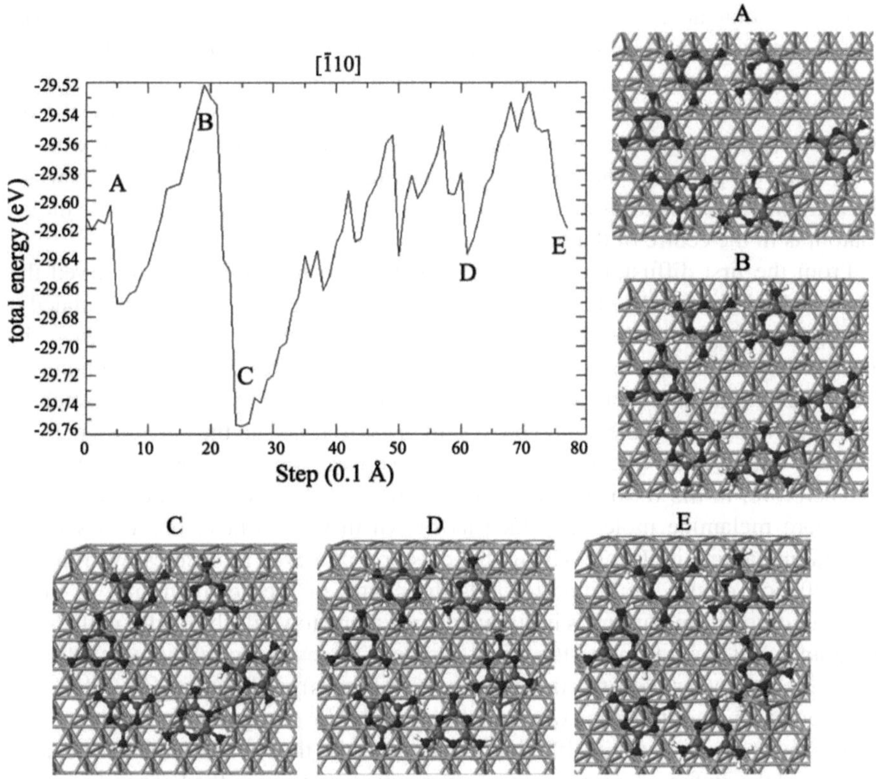

Fig. 5.13 The relative energy of a melamine molecule diffusing towards the Au ad-atom of the pentagon on the Au(111) surface calculated using the DFT method with the semi-empirical dispersion correction as a function of the molecule displacement towards the gap in the pentagon. Obtained geometries corresponding to some positions on the energy curve are also shown

Eventually, it appears to be outside the hexagon (geometry E). Thus, we found that the sixth melamine can pick up the Au atom from another melamine and close the hexagon pushing the Au atom outside it. Overall, Model B requires a number of energy barriers to overcome; however, neither of them is higher than 0.2 eV, which must be feasible at RT.

The second geometry considered corresponds to Model A in Fig. 5.11, with the starting geometry for our diffusion calculation and the direction in which the sixth melamine was moved shown in Fig. 5.12b. Unfortunately, this calculation results in the de-adsorption of the melamine from the surface. The result obtained underlines that the chosen direction for the sixth molecule does not lie close to the minimum energy path and hence we need to explore the potential energy surface (PES) of the system in more detail in order to better understand the process. It might still be possible that there is a relatively small barrier for the Au ad-atom to move from the side of the melamine to the centre of the hexagon and this barrier could be easily overcome at RT. After all, when the sixth melamine closes the hexagon, one should expect a considerable energy gain. Due to the considerable size of the system, we initially adopted a simple approach in which, starting from the configuration in Fig. 5.12b, we considered two cross-sections of the PES: one where the Au ad-atom is translated towards the centre of the hexagon from its in-block position with the sixth molecule pinned to the surface by its remote N atom, and the second one where the sixth melamine is translated towards the gap to close the hexagon, while the Au adatom is in the centre of the hexagon.

From the first diffusion calculation (pinned sixth melamine) we observed that when the gold is translated towards the centre, the melamine which it is bonded to is dragged along with the corresponding distortion of the hexagon. However, this process appears not to be very favourable because it makes the hydrogen bonds of the first melamine molecule with the others weaker: eventually the bond between the gold atom and the first melamine breaks. In order to translate the Au ad-atom to the centre of the hexagon one has to overcome the energy barrier of about 0.5 eV.

Therefore, in the following set of calculations, we considered the system with one more melamine molecule added as shown in Fig. 5.14a in order to stabilise the first melamine molecule. Using this cluster with seven molecules, we repeated the diffusion of the Au atoms towards the centre of the hexagon and, as in the previous case, the melamine is initially dragged along towards the centre with the Au ad-atom, but it then returns to the initial position. The energy profile of this diffusion calculation together with the three geometries (insets) is shown in Fig. 5.14a. The corresponding energy barrier was found to be 0.4 eV.

The last geometry of this diffusion calculation with the Au adatom in the hexagon centre and the sixth melamine molecule just outside the gap in the hexagon was then used as the starting geometry for the second diffusion calculation in which the sixth melamine was translated to close the gap in the hexagon, see Fig. 5.14b. We find no barrier along this diffusion path; moreover, there is a considerable energy gain of around 1.0 eV for the sixth melamine molecule to close the hexagon.

In summary, there is a barrier of about 0.4 eV for the gold atom to diffuse towards the centre of the hexagon, while there is a significant energy gain of 1.0 eV for the sixth melamine to close the hexagon on the Au(111) surface. These results indicate that the whole complicated process of the formation of a hexagon with a Au atom inside it due to Model A of Fig. 5.11, may actually be spontaneous (no barrier) or require a small barrier which can easily be overcome at RT.

Proper calculation of the energy barriers for either of the three models considered above would require more detailed calculations based on e.g. NEB method [11–14]. However, due to lack of time, these calculations (which are currently being run) are not included in this thesis.

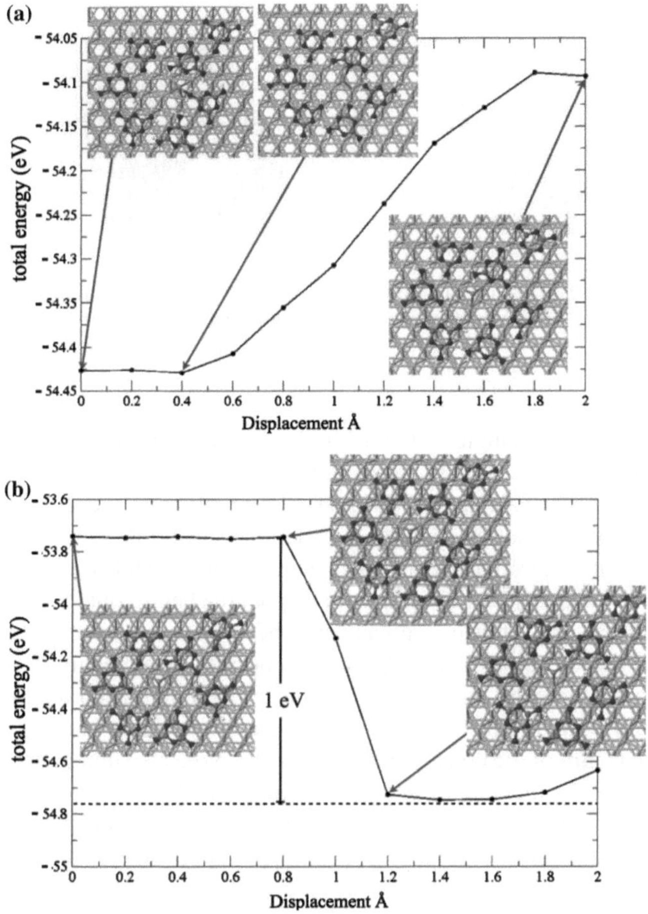

Fig. 5.14 The total energy of the system in the diffusion calculations shown as a function of the displacement coordinate: **a** when the Au ad-atom moves towards the centre of the hexagon and **b** when the melamine molecule moves to close the hexagon with the Au ad-atom being in its centre

5.7 Modelling the STM Images of the Hexagonal Cluster on the Au(111) Surface

An essential additional point for the justification of the "trapped gold atoms" model to explain the observed bright spots in the melamine hexagonal network on the Au(111) surface is to model its STM image. Hence, we performed modelling of STM images for a number of structures using the Tersoff-Hamann method as explained in Sect. 2.9.1. In all our calculations described below the bias was set to −0.5 V (occupied states). Two layers of gold were used in all cases.

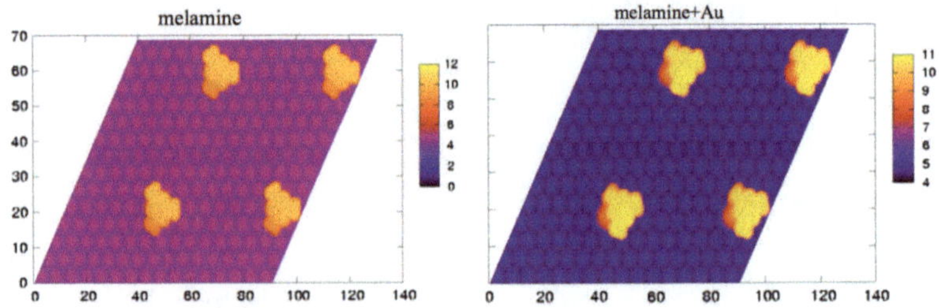

Fig. 5.15 Calculated STM images of a melamine with (*right*) and without (*left*) an Au ad-atom ($V_s = -0.5\text{V}, d_st = 3.6$). In the former case the system geometry corresponded to that shown in Fig. 5.4a

First, we analysed a single melamine molecule and a melamine molecule bonded to the Au ad-atom on the Au(111) surface. The geometry shown in Fig. 5.4a was used in the latter case. The calculated STM images for both cases are shown in Fig. 5.15. One can clearly see that an individual melamine molecule appears as a triangle, in agreement with the available experimental information, while the block "melamine + Au ad-atom" appears as a triangle with a spot attached to one of its sides. As far as we are aware, this latter image has never been observed experimentally, which may suggest that this block may only exist temporarily as a precursor for the growth of the hexagonal network with the gold atoms trapped in it: according to our model described above, as the hexagon is formed, the Au ad-atom leaves its block and moves to the centre of the corresponding hexagon. Note also that in both images the amino groups of the melamine are easily distinguishable.

The next system we considered was the hexagon of melamine molecules without and with the Au atom in its centre. The corresponding calculated STM images are shown in Fig. 5.16. All the geometries that we considered have been described in the previous sections. One can see from Fig. 5.16 (left) that it is possible to resolve every single melamine within the hexagon in the image. When the Au ad-atom is present in the hexagon centre, see Fig. 5.16 (right), there is a clear bright spot appearing in the STM image in the centre of the hexagon and the density of states of the single molecules is slightly perturbed due to a distortion in the geometry of the hexagon. In this case the gold atom is at the same height as the hexagon itself and it is equally bright.

Hence, we see that the expected appearance of the gold atom in the centre of a melamine hexagon is in perfect agreement with the bright spots observed in the STM images in Fig. 5.1a. This result brings in an additional justification to our model presenting the observed bright spots as being gold atoms trapped in the hexagonal network during its formation.

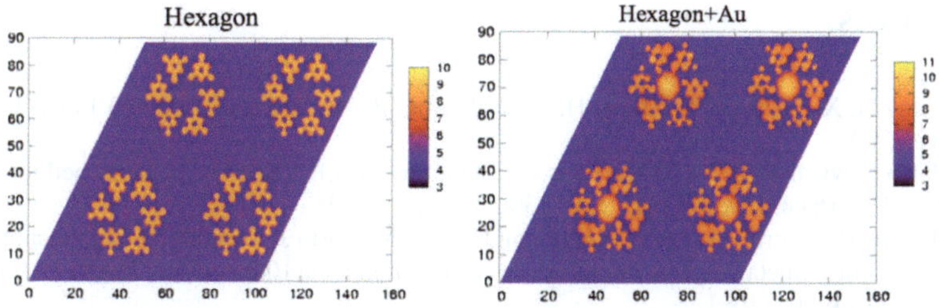

Fig. 5.16 Calculated STM images of the melamine hexagon without (*left*) and with (*right*) the Au ad-atom in its *centre* on the Au(111) surface ($V_s = -0.5V$, $d_st = 3.6$)

5.8 Conclusion

In order to prove our hypothesis that gold atoms are trapped in the melamine hexagonal network we used different approaches. We find a small diffusion barrier for a gold atom on the Au(111) surface. Once a "melamine +Au ad-atom" block is formed, they diffuse easily on the surface.

The diffusion calculations show that if a melamine approaches a "melamine+Au ad-atom" block from a direction different to the position of the Au atom, a hexagon may eventually form at RT with the Au atom either inside or outside it. Even in the case of the Au atom appearing outside the hexagon, when additional molecules come to form another hexagon attached to the first one, this Au atom would end up in the centre of the second hexagon. This observation also agrees with the experimental image in Fig. 5.1a where one can see gaps inside some of the hexagons, i.e. the space inside hexagons is empty (a "defect").

Summarising, we believe that, at small deposition rate, gold atoms, which are always present on the surface and quite mobile on it, would quickly form blocks or complexes with the melamine molecules, and these can also diffuse freely on the surface. When a second melamine molecule comes in proximity of these blocks, a hexagon formation is initiated. If the number of gold atoms is sufficiently high, then hexagons with the gold atoms trapped inside them are formed and the hexagonal network grows with the gold atoms inside almost every hexagon. However, if the deposition rate is high, then the melamine network is formed first as the gold atom would not have enough time to reach the molecules before this happens. Of course, some of them may, and occasional bright spots in the STM image in Fig. 5.1b taken at high deposition rate prove that.

A proper "proof" of this mechanism can only be done, based on the barriers (and hence rates) calculated in this study, if a dynamical approach is adopted, e.g. by means of an appropriate Kinetic Monte Carlo (KMC) [7] calculation which we plan as future work.

References

1. Zhang H, Xie Z, Long L, Zhong H, Zhao W, Mao B, Xu X, Zhao W (2008) J Phys Chem C 112:4209
2. Maksymovych P, Jr T (2008) Yates. Au adatoms in self-assembly of benzenethiol on the au(111) surface. J. Am. Chem. Soc. 130(24):7518–7519
3. Messina P, Dmitriev A, Lin N, Spillmann H, Abel M, Barth J, Kern K (2002) Direct observation of chiral metal-organic complexes assembled on a cu(100) surface. J. Am. Chem. Soc. 124(47):14000–14001
4. Böhringer M, Schneider W-D, Glöckler K, Umbach E, Berndt R (1998) Adsorption site determination of ptcda on ag(110) by manipulation of adatoms. Surf Sci Lett 419:L95–L99
5. Staniec PA, Perdigão LMA, Rogers BL, Champness NR, Beton PH (2007) J. Phys. Chem. C 111:886
6. Perdigão LMA, Perkins EW, Ma J, Staniec PA, Rogers BL, Champness NR, Beton PH (2006) J. Phys. Chem. B 110:12539
7. Gillespie DT (2007) Stochasti simulation of chemical kinetics. Annu. Rev. Phys. Chem. 58:35–55
8. Boisvert Ghyslain, Lewis Laurent J, Puska Martti J, Nieminen Risto M (1995) Energetics of diffusion on the (100) and (111) surfaces of ag, au, and ir from first principles. Phys. Rev. B 52(12):9078–9085
9. Bulou H, Massobrio C (2005) Mechanisms of exchange diffusion on fcc(111) transition metal surfaces. Phys Rev B 72(20):205427
10. Piana S, Bilic A (2006) J. Phys. Chem. B 110:23467
11. Mills G, Jonsson H (1994) Phys. Rev. Lett. 72:1124
12. Mills G, Jonsson H, Schenter G (1995) Surf. Science 324:305
13. Henkelman G, Uberuaga BP, Jonsson H (2000) J. Chem. Phys. 113(22):9901–9904
14. Henkelman G, Jonsson H (2000) J. Chem. Phys. 113(22):9978–9985

Chapter 6
Modelling of DNA Derivatives and Comparison with Experimental Results

In this chapter we address the characterisation of the assembly of N-aryl-modified nucleobases adenine, cytosine, guanine and thymine (DNA derivatives) on the Au(111) surface. The Watson-Crick hydrogen-bonding has traditionally been thought to be a prerequisite for high-fidelity DNA replication [1], playing an important role in the prebiotic soup and in the formation of organic systems there. At the beginning, surfaces may have mediated in the recognition events, catalyzing the formation of a covalent backbone prototype and may have functioned as templates in a primitive copying mechanism. However, considering the deposition of DNA and RNA bases alone on the Au(111) surface, Watson-Crick pairs seem not to be the only possibility for the assembly of these nucleotides [2–16].

In particular we shall consider simple nucleotides in which the nitrogen atom, normally attached to the sugar moiety, has been connected to an aromatic group (a kind of phenyl ring) serving as a substitute to the sugar phosphate group,[1] see Fig. 6.1. These modifications of the molecular structure of the nucleotides must play an important role in the recognition between complementary nucleobases as the possibility to form H-bonded structures is reduced because of the presence of the aromatic groups bonded to the nucleobases. In experiments conducted in UHV by our collaborators from the Aarhus University the DNA derivatives were deposited at a temperature close to 400 K on the Au(111) surface. The STM images were recorded in constant current mode and at temperatures ranging between 100 and 160 K, in order to stabilise the formed supramolecular nanostructures. They found that these molecules form well ordered patterns, see Fig. 6.2.

The modified cytosine (C) and guanine (G) molecules form monolayers consisting of parallel lines of molecules. However, the mixed phase of C and G shown in Fig. 6.2 is a porous network with six molecules in the unit cell. The adenine derivative, similar to guanine and cytosine, form monolayers consisting of parallel lines of molecules. The thymine derivative, on the other hand, does not form a periodic network.

[1] Details about the synthesis and the characterisation of the N-aryl-modified nucleobases could be found in [17, 18].

M. Mura, *Self-Assembly of Flat Organic Molecules on Metal Surfaces*,
Springer Theses, DOI: 10.1007/978-3-642-30325-8_6,

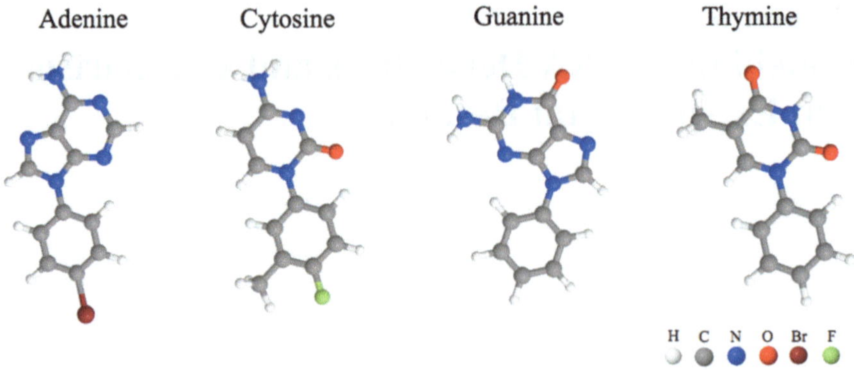

Fig. 6.1 Geometries of the DNA derivative: adenine, cytosine, guanine and thymine

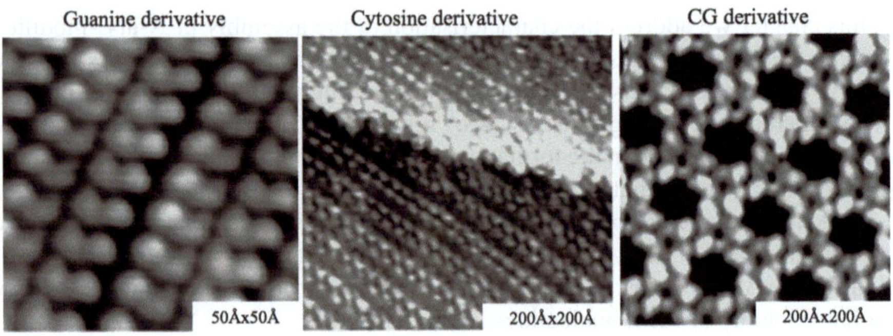

Fig. 6.2 High-resolution STM images of self-assembled nanostructures formed by depositing N-aryl-modified nucleobases guanine, cytosine and guanine-cytosine (from *left* to *right*) on the Au(111) surface

No pictures of these networks are given since we are mainly concerned with C and G network here.

The STM images of the DNA derivatives demonstrate different assemblies to those seen for non modified bases deposited on the same Au(111) surface [2, 9, 11, 15, 19]: the presence of the aromatic ring bonded to the DNA base makes different configurations in the assembly of the DNA bases more favourable. These experiments also indicate that the Watson-Crick hydrogen bonding can still play an important role in driving the molecular recognition between the complementary nucleobases on the Au(111) surface. Our aim here is to work out the molecular models responsible for these patterns. From the images it is not immediately clear how the molecules assemble. In the case of the mixed cytosine-guanine phase it is especially difficult to identify the molecules and distinguish between them. The gas phase model, in first approximation, is a good approach to describe all possible assemblies of the molecules and to identify the monolayers based on the DNA derivatives. The gas phase model is also a good approximation in the case of the gold surface because its PES could be considered flat for some organic molecules such as melamine, PTCDA,

PTCDI, NTCDI and NTCDA (see Chap. 4) and the DNA bases [2, 9–11, 13, 15]. We believe this is also true for the DNA derivatives.

We could have considered all possible structures using our systematic approach (as explained in Sect. 3.1), and then compare the obtained 2D structures with the experimental images. However, since a large number of monolayers could be built we selected only those based on the most stable dimers exploiting the previous work done for simple DNA bases [3–6] and considering only the structures which could explain the periodicity of the STM images observed. Moreover, we want to keep a continuity between the previous and the present work, therefore, in order to compare the DNA derivative dimers with the simple DNA dimers, we kept the same nomenclature as used previously.

In the next section we shall present our results for the selected dimers based on the DNA molecules in the gas phase. We shall first examine the most favourable homo-pairs, and then study the guanine-cytosine hetero-pairs (including the Watson-Crick pair). We demonstrated in Sect. 3.7 that the PBE method is accurate enough to describe the molecule-molecule interaction. Therefore, all the gas phase structures have been relaxed using the PBE method, while we used the vdW-DF method to describe the molecule-surface systems.

6.1 Pairs Based on the DNA Derivatives in the Gas Phase

This section is dedicated to the DNA derivatives dimers in the gas phase. The results are presented starting from the adenine dimers and then cytosine, guanine, thymine dimers. Finally the hetero-pairs obtained combining the cytosine and guanine will follow.

Adenine Dimer

Adenine is an asymmetric molecule, in the previous work that concerns the gas phase analysis of pairs based on the simple DNA base adenine [5] it was demonstrated that there are 21 possible planar adenine pairs. In our case the number of possibilities is decreased due to the presence of the aromatic ring bonded to the adenine nucleobase. However, in spite of the presence of the aromatic ring, the number of pairs possible is still large. We considered only seven possible adenine dimers shown in Fig. 6.3, which are selected considering the most stable configurations obtained in [5]. Two additional pairs are also considered which involved hydrogen bonds stabilised by bromine, specific for this molecule.

The seven dimers are found to relax into stable geometries, however, after relaxation the aromatic groups of the adenine molecules are slightly rotated with respect to the plane of the nucleobases. The stabilisation energies associated with these structures are presented in Table 6.1 and they are in good agreement with the values previously obtained in the literature for adenine [5].

The A1A2 dimer is the most stable structure while the ABr1 and ABr2 dimers are the least stable. This is also evidenced by the electron density difference plots

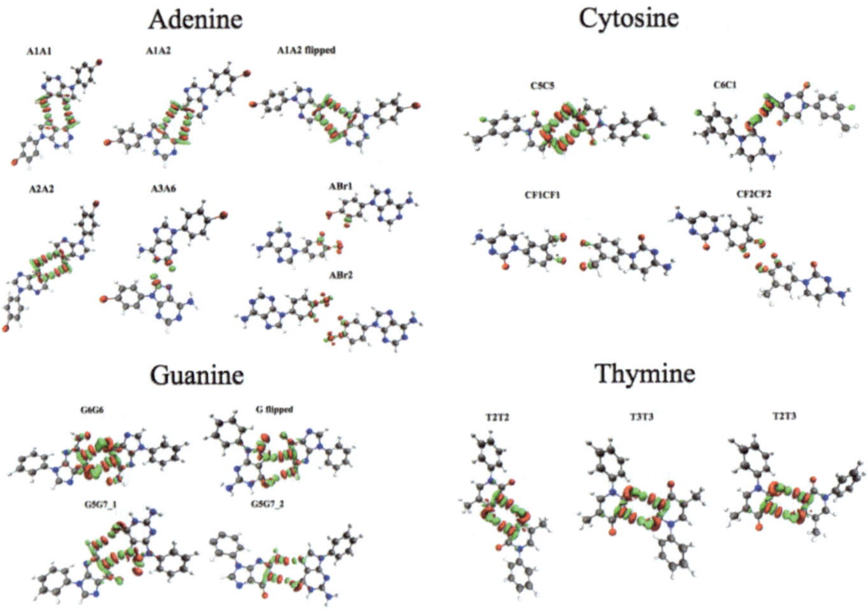

Fig. 6.3 The geometries of adenine, cytosine, guanine and thymine dimers shown together with the plots of the electron density difference corresponding to ±0.01 electrons/Å^3 for the A1A1, A1A2, A1A2_flipped dimer, C5C5, C6C1, the guanine and thymine pairs and ±0.002 electrons/Å^3 for ABr1, ABr2, CF1CF1, CF2CF2 pairs. The *green surfaces* correspond to the regions of positive electron density difference (excess) and the *red areas* correspond to the regions of negative electrons density difference (depletion)

Table 6.1 Energetics of the adenine derivative dimers (in eV)

	A1A1	A1A2	A1A2_flipped	A2A2	A3A6	ABr1	ABr2
E_{stab}	−0.46	−0.62	−0.48	−0.59	−0.14	−0.06	−0.03
E_{def}	0.09	0.10	0.16	0.11	0.003	0.01	0.01
E_{int}	−0.55	0.72	−0.64	−0.70	−0.14	−0.05	−0.04
E_{BSSE}	0.14	0.06	0.14	0.14	0.06	0.07	0.02

The stabilisation and interaction energies include the BSSE correction

in Fig. 6.3. The most developed “kebab” structure is displayed for the most stable dimers A1A2, A1A2, A1A2_flipped and A2A2 while in the case of the weaker pairs the “kebab” structures are interrupted. One may think that bromine could be a good acceptor in the hydrogen bond due to its high electronegativity (for bromine it is 2.96 eV compared with 3.04 and 3.44 eV for nitrogen and oxygen atoms respectively). However, the pairs are still weakly bound. The weak reactivity of the bromine is caused by the bond with the carbon atom of the aromatic ring.

So far the most suitable pairs that can be used to build the 1D chain are A1A2 and A2A2: the high stability of these pairs ensures a subsequent high stability when the molecules are deposited on the gold surface as indeed found in the experiments.

Table 6.2 Energetics of the cytosine and guanine derivatives dimers (in eV)

	Cytosine				Guanine			
	C5C5	C6C1	CF1CF1	CF2CF2	G6G6	G_flipped	G5G7_1	G5G7_2
E_{stab}	0.85	−0.36	−0.008	−0.001	−1.14	−0.79	−0.77	−0.56
E_{def}	0.18	0.08	0.018	0.015	0.21	0.10	0.10	0.04
E_{int}	−1.03	−0.44	−0.026	−0.017	−1.35	−0.89	−0.87	−0.59
E_{BSSE}	0.15	0.10	0.033	0.035	0.21	0.17	0.035	0.033

The stabilisation and interaction energies include the BSSE correction

Cytosine Dimer

Considering the simple cytosine DNA base, there are 13 cytosine homo-pair possibilities [3]. In the case of the derivative we considered only four possible dimers, shown in Fig. 6.3. Two of these pairs are stabilised by the same kind of contact as the simple DNA molecules considered previously in the literature [3], and there are also two more pairs that involve the fluorine atom in the hydrogen bond formation.

The dimers are found to relax into stable geometries, however, also in this case the plane of the aromatic ring is tilted with respect to the plane of the nucleobases. The stabilisation energies for these structures are presented in Table 6.2 and the values for C5C5 and C5C1 are in good agreement with the energies obtained for non-modified nucleotides [3]. The most stable dimer is, as expected, the C5C5 while the pairs that involve the fluorine atom (CF1CF1 and CF2CF2) are weak. The "kebab" structure of the C5C5 pair, in Fig. 6.3, is well defined showing the strength of the hydrogen bonds. This is in contrast with the "kebab" structures of the other dimers.

The dimer C5C5 therefore is a good candidate for building 1D chains.

Guanine Dimer

Considering the dimers based on all possible connections between the simple DNA guanine bases, there are 17 guanine homo-pairs [3]. In the case of the DNA guanine derivative the number of possibilities decreases and we considered only four of the possible guanine dimers which are shown in Fig. 6.3. These structures are similar to the simple DNA bases present in the literature [3] and we kept the same nomenclature used previously.

The dimers are found to relax into stable geometries in which the aromatic ring is tilted with respect to the plane of the nucleobases. The stabilisation energies of these structures are presented in Table 6.2 and they are in good agreement with the values obtained in the literature for simple guanine dimers [3]. The most stable dimer, as expected, is the G6G6 pair, however, all relaxed dimers have relatively high energies with respect to the DNA derivatives pairs considered so far. As a result, all "kebab" structures in Fig. 6.3 are seen to be well developed with respect to the weaker dimers described in the previous sections.

The G6G6 is the most favourable dimer, with the G_flipped and G5G7_1 being good candidates for the formation of 1D chains based on the guanine DNA pairs.

Table 6.3 Energetics of the thymine derivative dimers (in eV)

	T2T2	T3T3	T2T3
E_{stab}	−0.63	−0.56	−0.60
E_{def}	0.06	0.001	0.01
E_{int}	−0.69	−0.56	−0.61
E_{BSSE}	0.18	0.17	0.18

The stabilisation and interaction energies include the BSSE correction

Thymine Dimer

Previous calculations done for the simple thymine DNA base [6] highlight that there are 10 possible thymine homopairs. Here we considered three pairs, shown in Fig. 6.3, that are similar to the simple thymine DNA bases.

The dimers are found to relax into stable geometries and the aromatic ring is rotated (tilted) with respect to the plane of the nucleobases. The stabilisation energies of these structures are presented in Table 6.3. All dimers have similar stabilisation energies and these energies are in good agreement with the values presented in the literature for the corresponding simple thymine dimers [6].

In the thymine case all the pairs that we considered are good candidates for two molecule unit cells to be used as building blocks for the observed thymine structures.

6.2 Pairs Based on the Guanine and Cytosine DNA Derivative Molecules in the Gas Phase

This section is dedicated to the hetero-pairs based on the cytosine and guanine DNA derivatives. We considered only two pairs based on the configurations that are similar to the most stable geometries of the hetero-pairs based on the simple DNA bases present in literature [20], see Fig. 6.4. We expect the most stable pair to be the Watson-Crick (C8G10 in Fig. 6.4) that should play an important role when cytosine and guanine DNA derivatives are deposited on the surface.

From our calculations the most stable hetero-pair is the C8G10 (the Watson-Crick pair). The relative stabilisation energies of these pairs are in agreement with the previously obtained values for the simple G, C molecules [20]. The C8G10 pair must then be important in the assembly of the 2D structures, as it would ensure that such structure is to be the most stable (Table 6.4).

The next step is to build up 1D chains based on the DNA derivatives pairs and investigate the stability of these structures. Subsequently, that 2D structures can be constructed and compared with experiment.

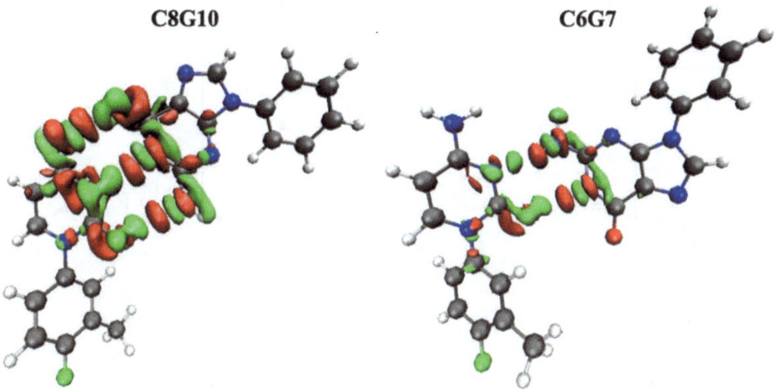

Fig. 6.4 Geometries of the guanine-cytosine dimers shown together with the plots of the electron density difference corresponding to ±0.01 electrons/Å^3. The *green surfaces* correspond to the regions of positive electron density difference (excess) and the *red areas* correspond to the regions of negative electrons density difference (depletion)

Table 6.4 Energetics of the cytosine-guanine derivatives dimers (in eV)

	C8G10	C6G7
E_{stab}	−1.25	−0.58
E_{def}	0.23	0.05
E_{int}	−1.48	−0.63
E_{BSSE}	0.24	0.15

The stabilisation and interaction energies include the BSSE correction

6.3 One-Dimensional Structures Based on the DNA Homo-Pairs in the Gas Phase

In the previous section we identified all most stable dimers that can be obtained by combining the binding sites available in the DNA derivatives. In this section, following our systematic method, we identified the free sites on the dimers that could be used to connect two dimers together into 1D structures. We shall present the results for all molecules keeping the same order as in the previous section.

Adenine

It is possible to derive all structures connecting the seven dimers we found. There are more than 10 possible chains, however, we account only for the chains that are stabilised by the most energetic connections between the pairs, see Fig. 6.5. Other cells could be obtained, if desired, by connecting the three most stable dimers through the weaker A3A6, ABr1 or ABr2 pair connections.

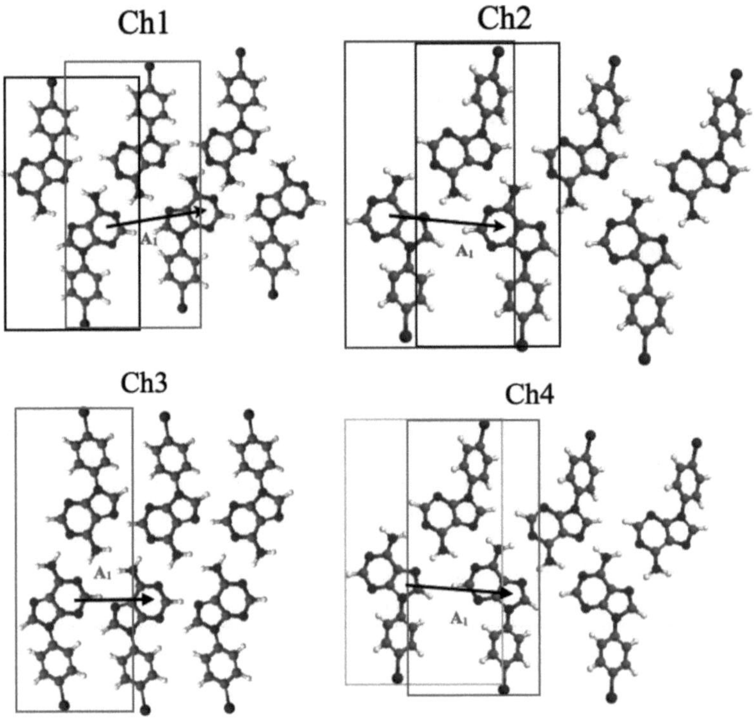

Fig. 6.5 Four possible 1D chains based on the adenine pairs. The unit cells and the lattice vectors are also shown. The *black*, *red*, *blue* and *green* unit cells, indicate, respectively, the A1A1, A2A2, A1A2_flipped and A1A2 pairs

Table 6.5 The evaluated (by summing up the corresponding dimer energies) stabilisation energies (in eV) for selected adenine chains

	Ch1	Ch2	Ch3	Ch4
E_{stab} (eval.)	−1.05	−0.94	−0.66	−1.07

The chains have not been relaxed and, to give an idea of the stabilities of these structures, we evaluated the energies of each unit cell by summing up the energies of the single contacts involved in the structures.

From our evaluations it results that the most stable chains are Ch1 and Ch4, see Table 6.5, which involve respectively the A1A1 and A2A2 dimers or the A2A2 and A1A2 dimers.

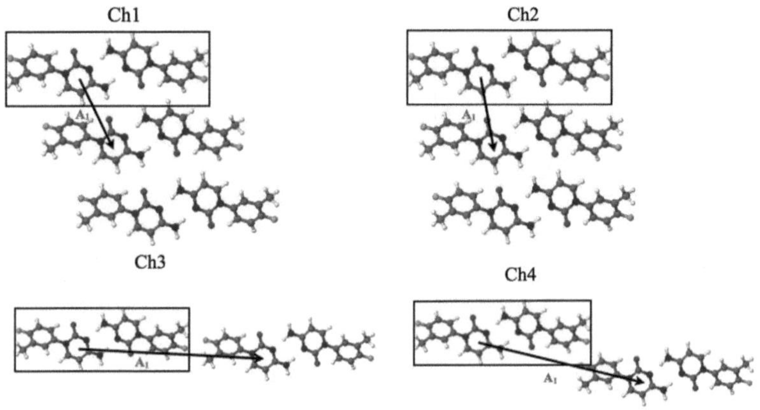

Fig. 6.6 Four possible 1D chains based on the C5C5 cytosine pair. The unit cell and the lattice vectors are also shown

Table 6.6 The evaluated (by summing up the corresponding dimer energies) stabilisation energies (in eV) for selected cytosine chains

	Ch1	Ch2	Ch3	Ch4
E_{stab} (eval.)	−1.55	−1.05	−0.85	−0.85

Cytosine

In the case of cytosine we considered only 1D chains based on the C5C5 pair as the unit cell. The most stable combination is Ch1 where the contacts between the dimers are stabilised by two strong hydrogen bonds, see Fig. 6.6. From our evaluation the most stable structures are Ch1 and Ch2, see Table 6.6.

The difference in energies between these chains is due to the dissimilarity in the type of contact between the pairs. In the case of Ch1 an oxygen atom of one pair and one hydrogen atom of the NH2 group of the second dimer are bound. In the Ch2 chain the contact is due to the oxygen and the hydrogen bonded to a CH group, which, as it has been shown for the PTCDA (see Sect. 3.2.1) does not produce strong hydrogen bonds.

Guanine

There are three stable 1D chains based on the guanine derivative pairs built form the chosen dimers: they are constructed combining together the G6G6 and the G5G7_1 pairs, see Fig. 6.7.

By summing up the energies of contacts of the guanine unit cell in the chain, we find that the most stable structure appear to be the Ch1 and Ch2, see Table 6.7, which

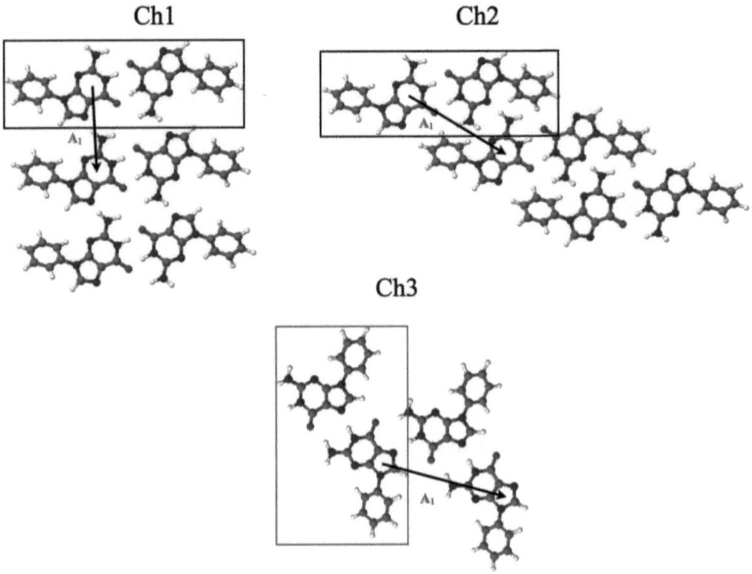

Fig. 6.7 Three possible 1D chains based on the guanine pairs. The unit cells and the lattice vectors are also shown. The *black and red* unit cells indicate, respectively, the G6G6 and G5G7_1 pairs

Table 6.7 Evaluated (by summing up the corresponding dimer energies) stabilisation energies (in eV) for selected Guanine chains

	Ch1	Ch2	Ch3
E_{stab} (eval.)	−1.58	−1.58	−1.16

have as the unit cell the G6G6 dimer; the contact between the dimers is stabilised by two N...H–N hydrogen bonds.

Thymine

The thymine 1D structures are obtained by considering as the unit cell the T2T3 and T3T3 pairs. The similarity between these pairs and the small number of binding sites available make it possible to obtain two 1D chains only, see Fig. 6.8.

We evaluated the stabilisation energies by summing up the single energies of contacts between the thymine pairs present in the chain. The most stable chain is Ch1 which involves the T2T2 and T2T3 dimers, see Table 6.8.

When all the chains are considered, the next step is to combine two similar chains in order to obtain the 2D structures, using the available binding sites to connect the chains together. The next section is devoted to presenting the 2D structures based on the DNA pairs.

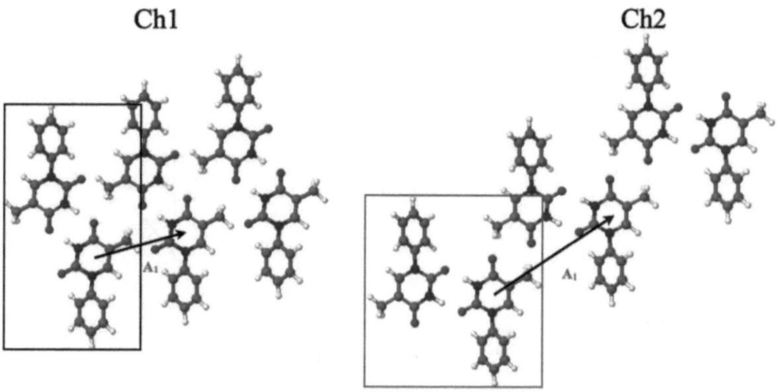

Fig. 6.8 Two possible 1D chains based on the thymine pairs. The unit cells and the lattice vectors are also shown. The *black and red* boxes indicate respectively the T2T2 and T3T3 pair as the corresponding unit cells

Table 6.8 The evaluated (by summing up the corresponding dimer energies) stabilisation energies (in eV) for selected thymine chains

	Ch1	Ch2
E_{stab} (eval.)	−1.23	−1.16

6.4 Gas-Phase Two-Dimensional Structures Based on the DNA Pairs

To build all possible 2D structures, we considered each chain separately and identified the binding sites which are available to connect two chains together. In all cases, the connections between the chains are possible only through weak interactions, because the strongest interactions are within the pairs or the chains. Therefore, we considered only two most stable 2D monolayers obtained by combining two similar 1D chains, for each DNA derivative.

Adenine

In order to analyse the 2D structures which can be formed by the adenine DNA derivative we combined together two Ch1 chains. We expect these monolayers to be very weakly bound because they involve weak ABr1 or ABr2 contacts, see Sect. 6.1. Only two monolayers have been considered and they are shown in Fig. 6.9. Other possible 2D monolayers could be obtained by connecting the less stable chains.

The most stable configuration is the Mon1 monolayer based on the Ch1 chain of the adenine DNA derivative, see Table 6.9.

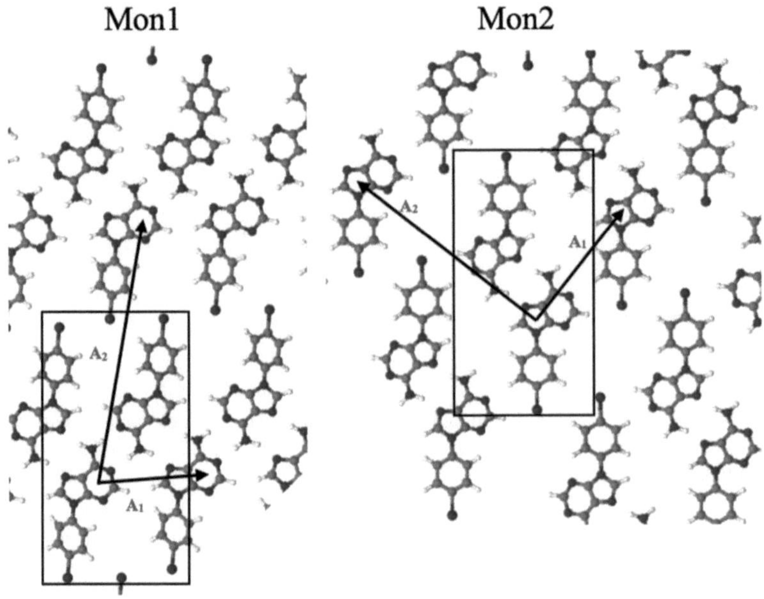

Fig. 6.9 Two possible 2D monolayers based on the Adenine pairs. The unit cells and the lattice vectors are also shown

Table 6.9 Energetics of the two adenine derivative monolayers (in eV)

	E_{stab}	E_{def}	E_{int}	E_{BSSE}	A_1	A_2	γ
Mon1	−0.91	0.15	−1.06	0.26	8.65	21.14	75.8
Mon2	−0.56	0.15	−0.70	0.22	10.82	18.16	91.4

The lengths of the lattice vectors (in Å) and the angle between them γ (in degrees) are also given. The stabilisation and interaction energies include the BSSE correction

Cytosine

In order to construct 2D structures due to the cytosine DNA derivative we again combined together two similar Ch1 chains. The connections between the cytosine chains are established by weak interactions due to the CF1CF1 and CF2CF2 contacts, see Fig. 6.10. Other possible monolayers could be obtained by connecting the less stable chains.

The most stable monolayer is Mon1 based on the Ch1 chain. The energies and the lattice vectors of the monolayers are shown in Table 6.10.

The monolayers are found to relax into stable geometries, however, the plane of the aromatic ring is tilted with respect to the plane of the nucleobases, as it is for the single dimers. This configurations of the dimers allowed monolayers to be stabilised by weak interactions due to the CF1CF1 and CF2CF2 contacts.

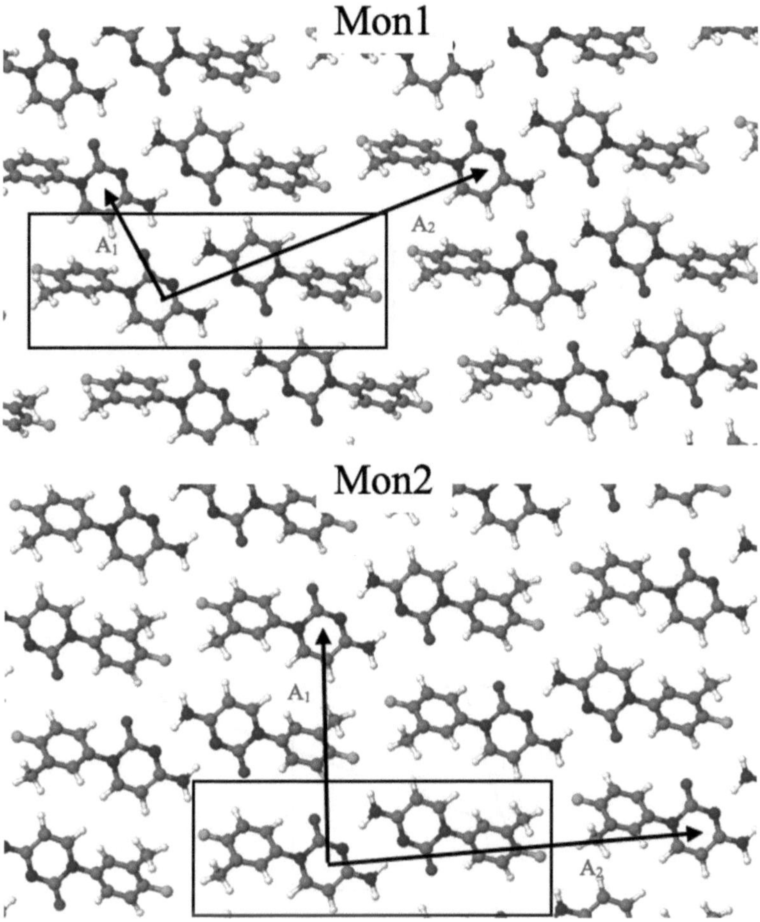

Fig. 6.10 Two possible 2D monolayers based on the cytosine pairs. The unit cells and the lattice vectors are also shown

Table 6.10 Energetics of the two cytosine derivative monolayers (in eV)

	E_{stab}	E_{def}	E_{int}	E_{BSSE}	A_1	A_2	γ
Mon1	−1.70	0.28	−1.98	0.44	6.87	19.86	93.4
Mon2	−1.01	0.18	−1.19	0.38	13.11	21.25	85.2

The lengths of lattice vectors (in Å) and the angle between them γ (in degrees) are also given. The stabilisation and interaction energies include the BSSE correction

Guanine

In the same way, we combined together the Ch1 and Ch3 chains that generate the monolayers Mon1 and Mon2 respectively, see Fig. 6.11.

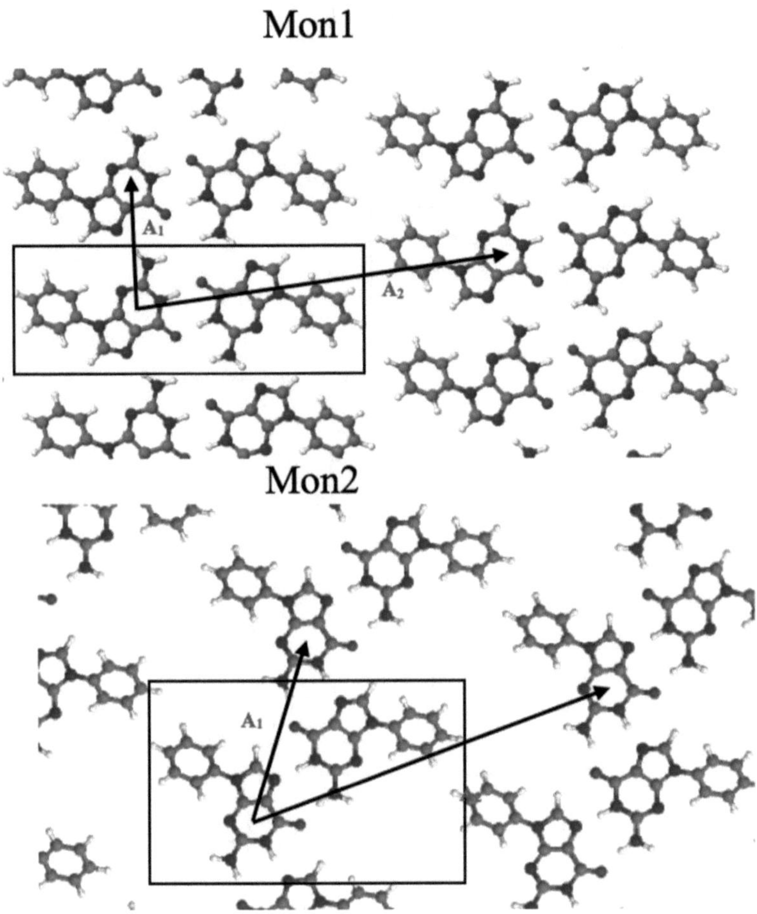

Fig. 6.11 Two possible 2D monolayers based on the guanine pairs. The unit cells and the lattice vectors are also shown

Table 6.11 Energetics of the two guanine derivative monolayers (in eV)

	E_{stab}	E_{def}	E_{int}	E_{BSSE}	A_1	A_2	γ
Mon1	−2.06	0.36	−2.42	0.36	7.38	22.8	104.7
Mon2	−1.86	0.39	−2.25	0.45	11.11	22.6	53.4

The lengths of lattice vectors (in Å) and the angle between them γ (in degrees) are also given. The stabilisation and interaction energies include the BSSE correction

The monolayer Mon1 is the most stable one, see Table 6.11, and it is a good candidate for the comparison with experimental data.

The modelling of these 2D structures is the starting point for the interpretation of the experimental STM images. The gas phase is a good approximation if the PES of the substrate is flat with respect to the molecules. In order to investigate in more detail the assembly of the molecules, we have to study the interaction between the DNA bases derivatives and the Au(111) surface. The next section is dedicated to the analysis of the molecule-surface interaction.

6.5 Interaction with the Au(111) Surface and STM Modelling

The gas phase investigation gives information on the assembly of the molecules, however, the Au(111) surface could play a role in the orientation of the molecules and eventually in their assembly. Especially, as in some cases, as was demonstrated above, molecules arrange themselves not necessarily in completely planar geometries. Therefore, we performed calculations in which we placed single molecules first and subsequently, a pair of them, on the gold surface. We positioned the molecules at a distance of 3.5 Å above a 4×5 gold cell, then we relaxed the geometry with our vdW-DF method. We also modeled the STM images of the relaxed DNA bases on the surface in order to be able to identify the molecules on the experimental STM images. In this section, our attention is focused only on the cytosine and the guanine DNA derivatives interacting with the Au(111) surface.

6.5.1 Single DNA Bases on the Au(111) Surface

We placed single cytosine and guanine DNA derivatives on a 4×5 gold cell to lie flat on the surface. After the relaxation with the vdW-DF method the guanine is still flat on the Au(111) surface while the cytosine does not stay flat, see Fig. 6.12. In the case of cytosine only the aromatic ring lies flat on the surface. The electronic density difference plot of the cytosine on the Au(111) surface provides evidence of the distortion of the nucleobase and of the interaction between the peripheral atoms of the molecule with the surface, see Fig. 6.12.

The adsorption energies for the relaxed configurations of the cytosine and guanine were found to be respectively -1.29 and -1.22 eV at around 3.4 and 3.2 Å separation from the surface (Table 6.12).

The relaxed geometry of the molecule on the surface is the starting point for the modeling of the STM image. Information coming from the calculated STM images of the single DNA base derivatives is necessary to identify the single molecules in the experimental images and work out the periodicity and composition of the network, when necessary.

In order to define the right value of intensity of the current to use in the modelling of the STM image we first evaluated the variation of the current intensity I(Z) with the distance Z between the tip and surface, see Fig. 6.13a, d. The images were calculated in the constant current mode using a W(110) tip whose apex was composed of five tungsten atoms with the average distance between tip and sample of 6.5 Å and the following scanning parameters: $V_s = -1.00$ V, $I_t = 10$ nA.

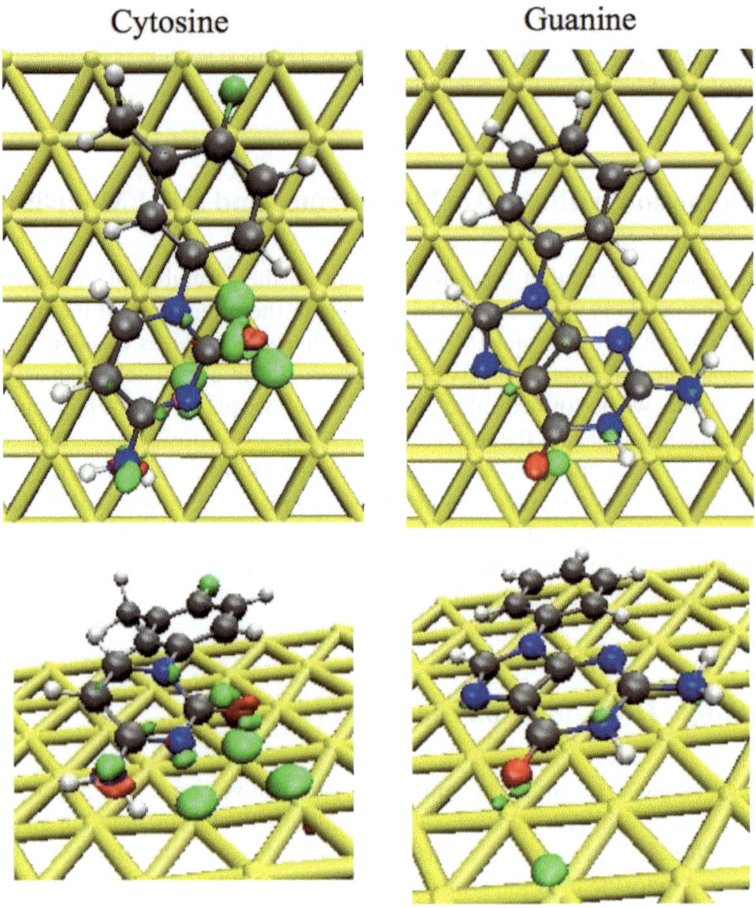

Fig. 6.12 The side (*bottom*) and top (*top*) views of the relaxed structures and the electronic density difference plots for the cytosine and guanine DNA derivatives on the Au(111) surface obtained by subtracting the densities of isolated molecules and of the surface, in the geometries of the combined system, from the density of the latter. The *red and green* contours correspond to depletion and excess (at ± 0.01 electrons/Å^3) of the electron density, respectively. Only the upper layer of the gold surface is shown for clarity

The images of the C and G molecules calculated using Nt_STM code, are shown in Fig. 6.13. They show similar elongated features in both cases.

Our models for the cytosine may not be able to describe the real case of the monolayer because the single nucleobase is tilted towards the surface while in the monolayers the nucleobase is bonded to the nucleobase of another DNA molecule and it may instead go flat with respect to the surface. For this reason, it would be more useful to consider the interaction between the pairs of DNA derivatives and the

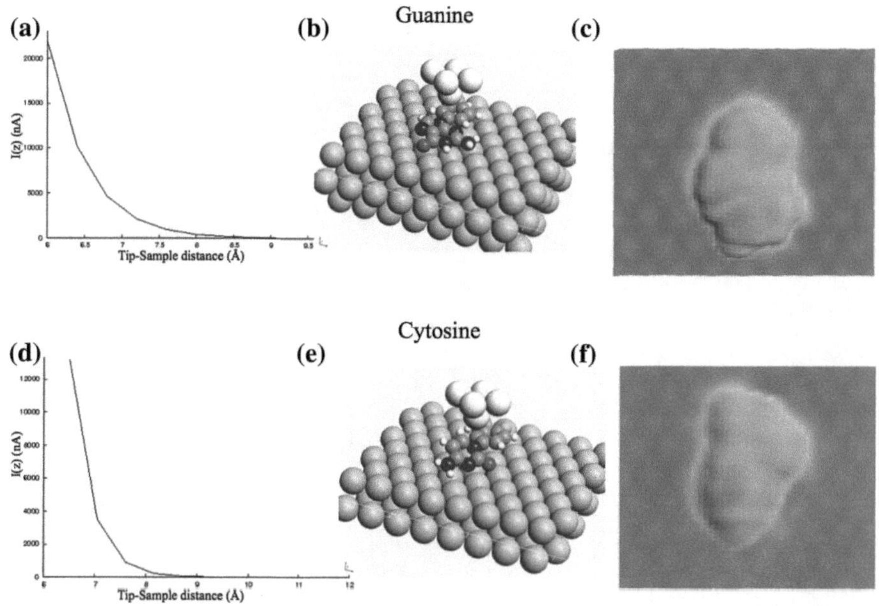

Fig. 6.13 Calculated profile of the STM intensity of current in function of the position of the tip I(z) (**a, d**) and STM images of (**c**) cytosine and (**f**) guanine molecules on the Au(111) surface using a W(110) tip ($V_s = -1.00\,\text{V}$, $I_t = 10\,\text{nA}$). The geometries (**b, e**) of the molecules in the presence of the W(110) tip apex are shown

Table 6.12 Energies for single DNA derivatives interacting with the Au(111) surface, studied using the vdW-DF method

	Cytosine	Guanine
Adsorption energy, eV	−1.29	−1.22
E_{def}	0.08	0.09
E_{int}	−1.37	−1.31
E_{BSSE}	0.31	0.31
Height above the surface, Å	3.4	3.2

Au(111) surface. In the next section we consider cytosine, guanine and the cytosine-guanine pairs interacting with the surface (Table 6.12).

6.5.2 DNA Base Pairs on the Au(111) Surface

We considered the C5C5 and the G6G6 homopairs and C8G10 heteropair on the Au(111) surface. In these calculations we placed the pairs 3.5 Å above a 4 × 5 gold cell. After the relaxation both the DNA derivatives lie flat on the surface while the

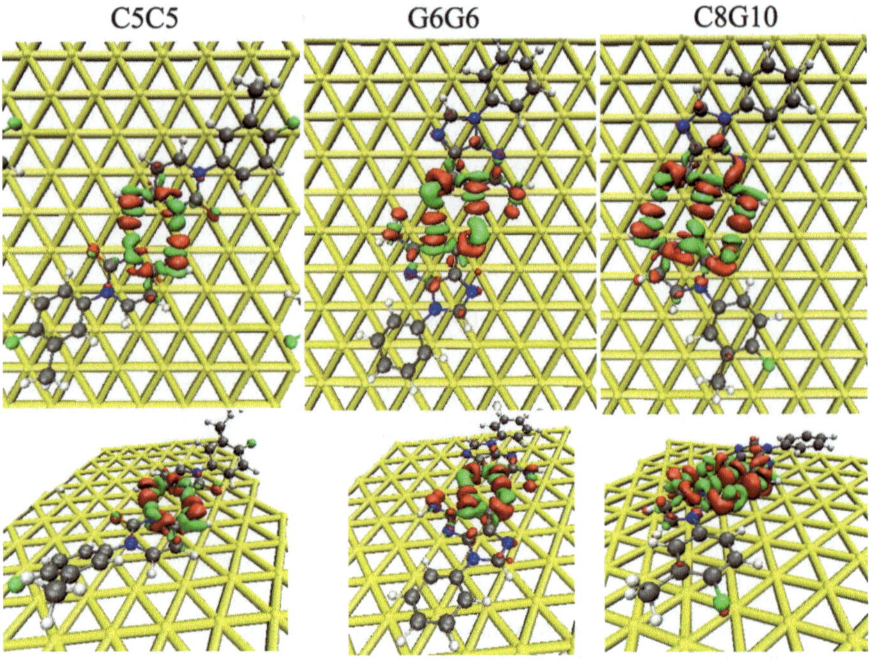

Fig. 6.14 The side (*bottom*) and top (*top*) views of the relaxed geometries and electronic density difference plots for the cytosine and guanine DNA derivative pairs and the cytosine-guanine heteropair on the Au(111) surface. The density difference was obtained by subtracting the densities of isolated molecules and of the surface, in the geometries of the combined system, from the density of the latter. The *red and green* contours correspond to depletion and excess (at ± 0.01 electrons/Å^3) of the electron density, respectively. Only the upper layer of the gold surface is shown

plane of the aromatic ring is slightly tilted with respect to the plane of the nucleobase of the DNA derivative, see Fig. 6.14. The calculation does not predict any change in the hydrogen bonding, see Fig. 6.14, and the overall binding energy of two single molecules to the gold is basically equal to the sum of the isolated dimer binding energy and twice the single molecule adsorption energy.

The calculated adsorption energies for the C5C5 and G6G6 were found to be respectively -1.98 and -2.39 eV, at the molecule-surface separations of around 3.5 and 3.6 Å, while for the C8G10 heteropair were found to be -2.32 eV, at the molecule-surface separations of 3.1 Å, see Table 6.13. We found that the binding energies are quite strong and the geometries of the dimers are not modified by the presence of the Au(111) surface. As a result, all "kebab" structures in Fig. 6.14 are seen to be well developed and are similar to those in the gas phase ones, as described in the previous Sects. 6.1 and 6.2.

The obtained relaxed geometries of the pairs were used to model their STM images. These images were calculated in the constant current mode using the W(110) tip composed of five tungsten atoms as its apex with the average tip-sample distance

Table 6.13 Energetic and the height above the surfaces of DNA derivative homopairs and heteropairs placed on the Au(111) surface and studied using the vdW-DF method

	Cytosine	Guanine	Cytosine-Guanine
Adsorption energy, eV	−1.89	−2.28	−1.99
E_{def}	0.002	0.02	0.33
E_{int}	−1.89	−2.31	−2.32
E_{BSSE}	0.43	0.57	0.49
Height above the surface, Å	3.5	3.6	3.1

of 6.5 Å and using: $V_s = -1.00$ V, $I_t = 10$ nA. The calculated STM images of pairs of molecules, shown in Fig. 6.15, provide a useful help for the correct interpretation of the experimental STM images.

All our calculations have been done in order to work out the molecular models which can reproduce the experimental data. The next section is dedicated to such a comparison with the experimental images.

6.6 Comparison with Experimental Data

Deposition on the gold surface of the guanine or cytosine DNA derivatives lead to well-ordered chains as shown in Fig. 6.2. In order to identify the correct structures to reproduce experiments, we first compared the experimental unit cells with those calculated with DFT.

In the case of the guanine the experimental unit cell ($a = 0.8 \pm 0.1$ nm, $b = 2.2 \pm 0.1$ nm and $\theta = 90° \pm 1°$) is indicated in Fig. 6.16. Initially we thought that the experimental STM image of the DNA guanine derivative could be identified with our monolayer Mon1, see Sect. 6.4. In Fig. 6.16 we compared the experimental STM image with the structures of Mon1 based on the G6G6 pair of the guanine derivative. The lattice vectors of our model are $a = 0.74$ nm, $b = 2.2$ nm and $\theta = 85.8°$, which are close to the experimental values.

In the case of the cytosine the experimental unit cell ($a = 0.7 \pm 0.1$, $b = 2.0 \pm 0.1$ nm and $\theta = 80° \pm 1°$) is also indicated in Fig. 6.16. It is possible to compare the experimental STM image of the cytosine derivative with our modelled monolayers Mon1 and Mon2, see Sect. 6.4. In Fig. 6.16 (bottom) we compared the STM image with the modelled 2D structure, Mon1, based on the C5C5 pair of the cytosine derivative for which the lattice vectors are: $a = 0.687$, $b = 1.986$ nm and $\theta = 93.4°$. The angle between the lattice vectors is slightly larger than in the experiment. This may be due to the lack of vdW interaction between the chains in our calculations.

In both cases, for cytosine and guanine, there is an agreement in the lengths of the lattice vectors but the angles between them are slightly incorrect. In the case of guanine a more detailed analysis shows that the Mon1 monolayer is not a good candidate to describe the real assembly of the guanine on the Au(111) surface. The

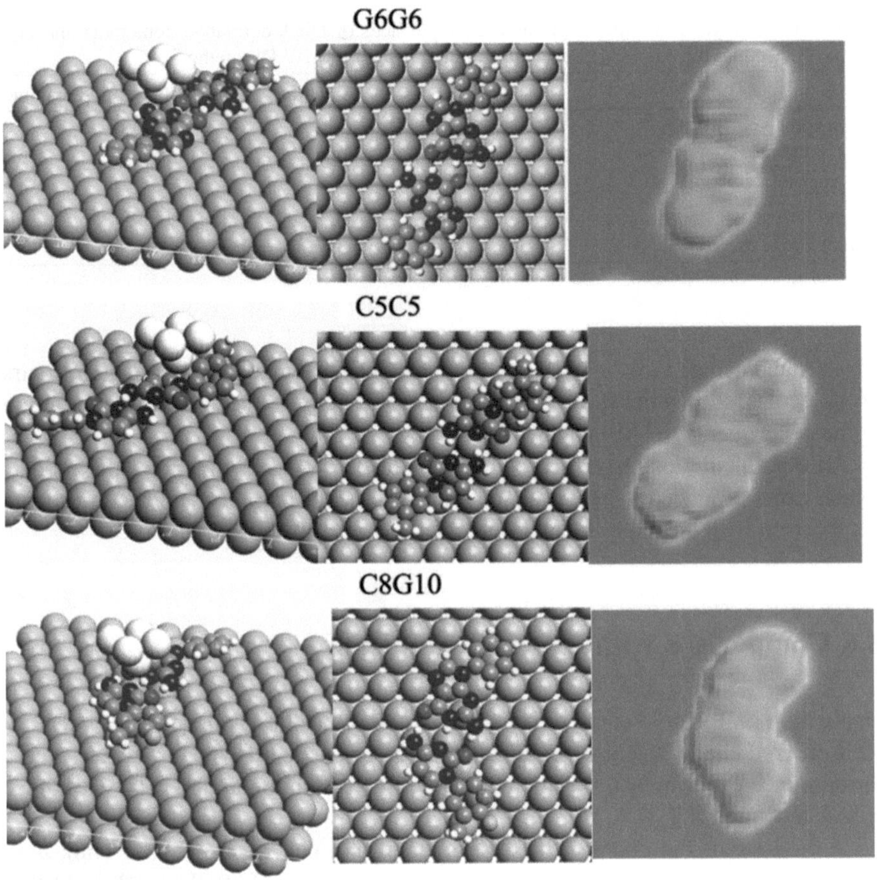

Fig. 6.15 STM modelling of the cytosine, guanine homopairs and cytosine-guanine heteropair on the Au(111) surface using the W(110) tip apex ($V_s = -1.00\,\text{V}$, $I_t = 10\,\text{nA}$)

molecules in our model form a zig zag shape and they are oriented along a different direction, however from the experimental STM image the molecules look oriented in the same directions. To evidence this aspect we have superimposed the calculated STM images of the guanine molecule on the experimental STM images as shown at the top of Fig. 6.16.

The correct model for the guanine, that we called Mon3, is suggested by our collaborators Jian-guo Wang[2] [21] and it is shown at the bottom of Fig. 6.17. Considering the monolayer Mon3, the stabilisation of this monolayer is due to the contacts between the single unit cells while the connection between the molecules within the

[2] Institute of Industrial Catalysis, College of Chemical Engineering and Material Science, Zhejiang University of Technology, Hangzhou, China

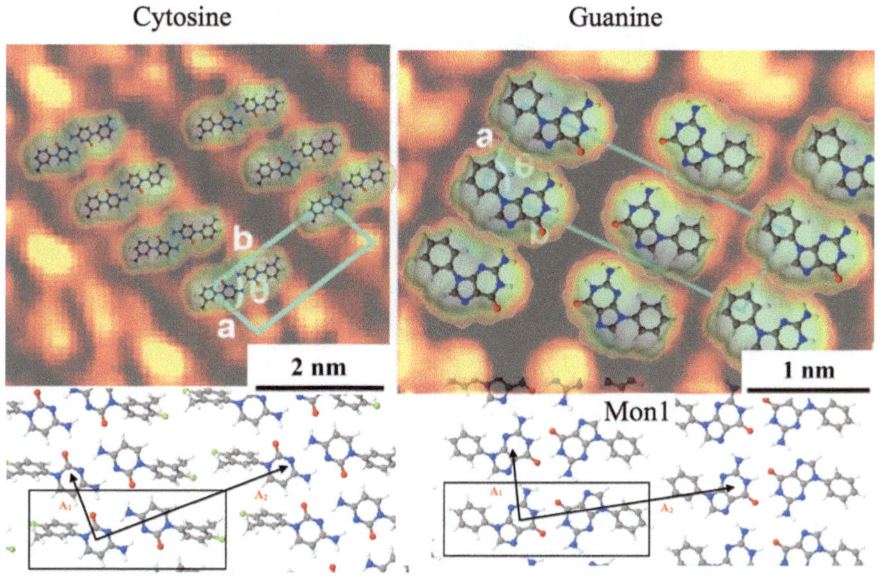

Fig. 6.16 High-resolution STM images of the self-assembled supramolecular network formed by the guanine and the cytosine derivatives. The calculated STM images of cytosine dimer C5C5 (*top left*) and of the guanine molecules (*top right*) are superimposed on the experimental images for clarity. At the bottom the DFT molecular models are presented together with the corresponding unit cells and the lattice vectors

unit cell is due to the interaction between one oxygen atom of one guanine and two hydrogens of the other guanine molecule.

We relaxed the monolayer Mon3 in the gas phase with our PBE method; the stabilisation energy is -2.3 eV, so that the Mon3 becomes the most stable amongst the monolayers considered. The lattice vectors of our model are $a = 0.75$ nm, $b = 2.1$ nm and $\theta = 90.3°$, which are in good agreement with the experimental values.

The obtained relaxed geometry of the guanine pair was used to model its STM image with the same parameters as in the previous calculations. The STM image of this pair of molecules is shown in Fig. 6.18. The model obtained has been superimposed on the experimental STM image in the top part of Fig. 6.17.

The analysis of the supramolecular network structure of the cytosine-guanine mixture, shown in Fig. 6.2, required a more detailed study. From a close up of the STM image in Fig. 6.19b it is possible to distinguish two kinds of molecules, however, it is very difficult to identify what these molecules are. What is known, is that in this case the Watson-Crick pair is the most stable: in gas phase, therefore, one could expect that the guanine and the cytosine are present in the sample in a ratio 1:1. Moreover, one could compare the STM image of the mixture with the single guanine and cytosine STM images in Fig. 6.13. From this comparison our collaborators in Aarhus recognised the guanine derivative as the triangle-linked-ball protrusions, see Fig. 6.19b, while the cytosines appear as the elongated shapes with subtle sub-features

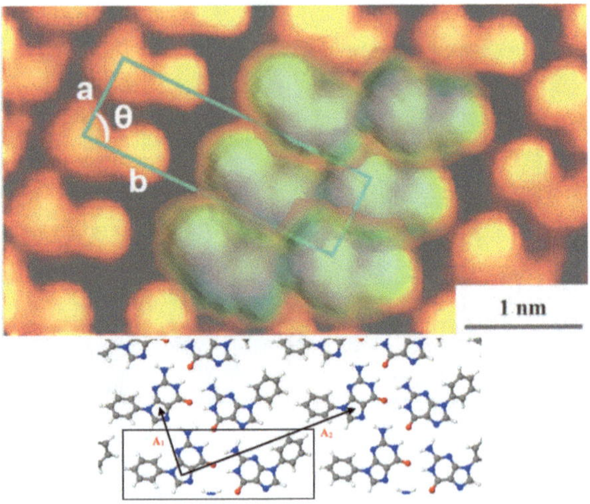

Fig. 6.17 High-resolution STM image (*top*) of the self-assembled supramolecular network formed by the guanine derivatives. At the *bottom*, the DFT molecular model is presented together with the corresponding unit cell and the lattice vectors. The calculated STM images of the G pair are superimposed on a part of the experimental image to demonstrate the validity of the model

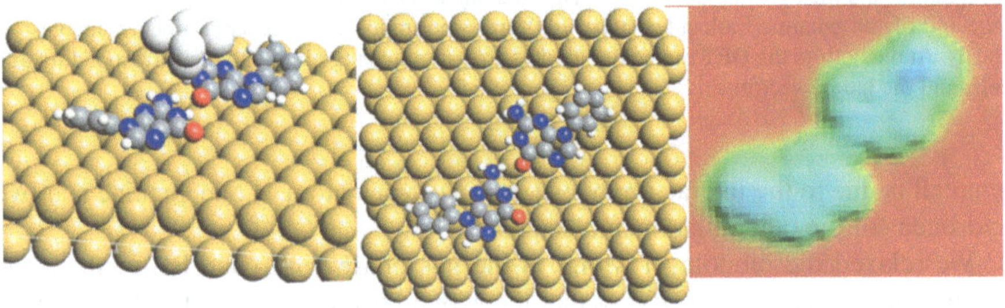

Fig. 6.18 STM modelling of the guanine homopair on the Au(111) surface involved in the guanine monolayer of Fig. 6.17, using the W(110) tip ($V_s = -1.00$ V, $I_t = 10$ nA)

in Fig. 6.19b. Having this information, it is possible to work out the periodicity of the network and its unit cell, which is composed of six molecules. The correct unit cell that could reproduce the supramolecular network is shown in Fig. 6.19d and it has been relaxed by Jian-guo Wang[3] [21]. The success of this model, shown in Fig. 6.19c, is demonstrated when we superimposed the calculated STM image of our cytosine-guanine pair on the STM image of the network, in which the shape and angle of the model match well with the experimental image.

The results of this study have recently been submitted [21].

---

[3] Institute of Industrial Catalysis, College of Chemical Engineering and Material Science, Zhejiang University of Technology, Hangzhou, China

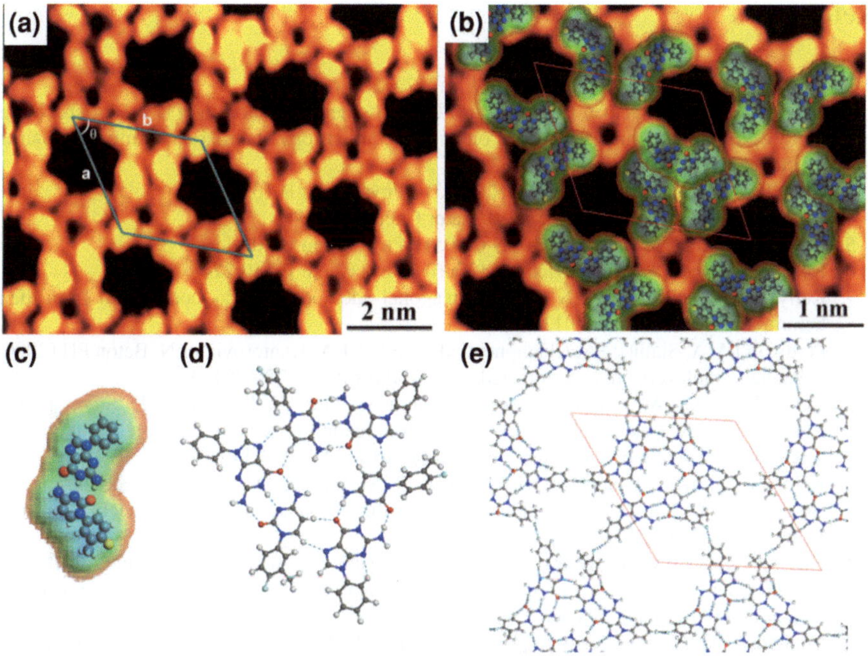

Fig. 6.19 **a** High-resolution STM images of the self-assembled supramolecular network formed by the cytosine and guanine molecules; **b** DFT calculated molecular models superimposed on the STM image in which the Watson-Crick CG pair are highlighted by the green ovals; **c** the theoretical geometry and STM model of the CG dimer and **d** the hexagonal unit cell used as a model for the whole network together with the corresponding unit cells and the lattice vectors (**e**)

6.7 Conclusion

In conclusion, the gas phase approximation was found to be a good approach to describe all possible assemblies of the molecules and identify the monolayers based on the DNA derivatives pairs. In order to work out the molecular periodicity of the networks we considered the interaction between the single molecules and the dimers of DNA derivatives with the Au(111) surface. Indeed, the geometries of the DNA molecules and pairs on the surface, calculated with the vdW-DF method, are similar to the gas phase ones and the "kebab" structures also do not change for the dimers on the surface as compared with the gas phase one. The strong binding energies give more proof of the significant influence of the vdW forces on the interaction between these flat organic molecules and the Au(111) surface.

The performed STM modelling of the DNA derivatives was successfully employed for the comparison with the experimental STM images. We have also shown that these models can provide a useful tool for the correct interpretation of the experimental STM images. Some of the results of this chapter have recently been submitted [21].

References

1. Watson JD, Crick FHC (1953) Nature 171:737–738
2. Xu REA, Kelly W, Gersen H, Lægsgaard E, Stensgaard I, Kantorovich LN, Besenbacher F (2009) Prochiral guanine adsorption on au(111): an entropy-stabilized intermixed guanine-quartet chiral structure. Small 5:1952
3. Kelly REA, Lee YJ, Kantorovich LN (2005) J Phys Chem B 109:22045
4. Kelly REA, Kantorovich LN (2006) J Mater Chem 16:1894
5. Kelly REA, Lee YJ, Kantorovich LN (2005) J Phys Chem B 109:11933
6. Kelly REA, Lee YJ, Kantorovich LN (2006) J Phys Chem B 110:2249
7. Kelly REA, Kantorovich LN (2005) Hexagonal adenine networks constructed from their homo-pairings. Surf Sci 589:139–152
8. Perdigão LMA, Staniec PA, Champness NR, Kelly REA, Kantorovich LN, Beton PH (2006) Adenine monolayers on ag-terminated si(111). Phys Rev B 73:195423
9. Kelly REA, Xu W, Lukas M, Otero R, Mura M, Lee YJ, Lægsgaard E, Stensgaard I, Kantorovich LN, Besenbacher F (2008) An investigation into the interactions between self-assembled adenine molecules and the au(111) surface. Small 4:1494
10. Kelly REA, Lukas M, Kantorovich LN, Otero R, Xu W, Mura M, Laesgaard E, Stensgaard I, Besenbacher F (2008) Understanding disorder of the dna base cytosine on the au(111) surface. J Chem Phys 129:184707
11. Otero R, Lukas M, Kelly REA, Xu W, Lægsgaard E, Stensgaard I, Kantorovich LN, Besenbacher F (2008) Elementary structural motifs in a random network of cytosine adsorbed on a gold(111) surface. Science 319:312–315
12. Mamdouh W, Dong M, Kelly REA, Kantorovich LN, Besenbacher F (2007) Coexistence of homochiral and heterochiral adenine domains at the liquid/solid interface. J Phys Chem B 111:12048–12052
13. Lukas M, Kelly R, Kantorovich L, Otero R, Laesgaard E, Stensgaard I, Besenbacher F (2009) Adenine monolayers on the au(111) surface: structure identification by stm experiment and ab initio calculations. J Chem Phys 130:024705
14. Mamdouh W, Kelly REA, Dong M, Kantorovich LN, Besenbacher F (2008) Two-dimensional supramolecular nanopatterns formed by the coadsorption of guanine and uracil at the liquid/solid interface. J Am Chem Soc 130:695–702
15. Xu W, Kelly REA, Schöck M, Otero R, Laesgaard E, Stensgaard I, Kantorovich LN, Besenbacher F (2007) Probing the hierarchy of thymine-thymine interactions in self-assembled structures by manipulation with scanning tunneling microscopy. Small 3:2011–2014
16. Otero R, Xu W, Lukas M, Kelly REA, Laegsgaard E, Stensgaard I, Kjems J, Kantorovich L, Besenbacher F (2008) Specificity of watson-crick base pairing on a solid surface studied at the atomic scale. Angew Chem Int Ed 47:9673–9676
17. Jacobsen MF, Andersen CS, Knudsen MM, Gothelf KV (2007) Org Lett 9:2851–2854
18. Jacobsen MF, Knudsen MM, Gothelf KV (2006) J Org Chem 71:9184–9190
19. Xu W, Dong M, Gersen H, Rauls E, Vázquez-Campos S, Crego-Calama M, Reinhoudt DN, Stensgaard I, Laegsgaard E, Linderoth TR, Besenbacher F (2007) Small 3:854
20. Kelly REA, Kantorovich LN (2007) J Phys Chem C 111:3883
21. Xu W, Wang J, Jacobsen MF, Mura M, Yu M, Kelly REA, Meng Q, Laegsgaard E, Kjems J, Linderoth TR, Kantorovich L, Gothelf KV, Besenbacher F (2010) Supramolecular porous network formed by molecular recognition between chemical modified nucleobases guanine and cytosine. Angewandte Chemie Int Ed 49(49):9373–9377

Chapter 7
Conclusions

The main aim of this thesis was to characterise the STM images of assemblies of flat organic molecules on the Au(111) metal surface and give useful information about their assembly mechanism. In order to describe in detail the assembly of the molecules, the study has been approached using various methodologies and considering different aspects of the process. Our method could be summarised in the following steps: firstly, consider all possible connections between molecules, which enable them to form periodic 1D and 2D structures in the gas phase; secondly, using an ab initio method, relax the structures which are expected to be amongst the most energetically favourable, and, thirdly, consider the effect of the surface on the structures geometry and mobility on the Au(111) surface and correspondingly adjust the model. In more difficult cases STM images are simulated to help in the characterisation.

The first part of the work presented in this thesis is concerned with the characterisation of the molecule–molecule interaction considering all possible connections between molecules, and the interaction with the surface has not been taken into account. Using a systematic methodology based on the identification of the binding sites on each molecule and applying well defined rules for connecting the molecules together, all possible super-structures can be constructed. In particular, considering a unit cell composed by two or more molecules, a large number of structures could be identified. These structures have been systematically relaxed with ab initio DFT methods. This data are indispensable for analysing and interpreting the structures formed on the surface. The effect of the vdW forces in the assembly of molecules on the surface has been considered thanks to a new ab initio method called vdW-DF.

The characterisation of the possible unit cells is a key point because their structure is necessary to predict larger structures based on the unit cell itself. This approach is useful for selecting the most stable configurations to relax with vdW-DF and hence to optimise the computational cost.

In the case of the melamine molecular assembly, this systematic approach has been applied to construct 1D and 2D planar periodic melamine assemblies. It has been shown that there are only two possible structures in the gas phase which have two molecules in the unit cell; both structures have been observed experimentally

M. Mura, *Self-Assembly of Flat Organic Molecules on Metal Surfaces*,
Springer Theses, DOI: 10.1007/978-3-642-30325-8_7,

in STM images [1–3]. If four molecules are considered to form a unit cell, then 6 stable different monolayers are possible in addition to another one which has been observed experimentally in [1]. Other monolayers are possible but they are less stable since some of the molecules are only two-fold coordinated. More structures can be envisaged if, when building the periodic structures, more than four molecules in the unit cell are accepted. The lattice vectors and the geometries of the theoretical structures are in good agreement with the ones experimentally observed.

The same theoretical approach in constructing 1D and 2D planar periodic structures has been applied to PTCDA and PTCDI assemblies. A large number of 1D and 2D structures for PTCDA and PTCDI molecules have been found. Calculations show that the chains and the monolayers based on PTCDA dimers are very similar in energy and in many cases also in geometry. Three of the obtained monolayers have been observed experimentally as the square [4–10], herringbone [6, 8–13] and brick wall [14, 7] phases. There is good agreement between the theoretical and experimental geometries of these super-structures.

In the same way, eight possible structures have been predicted for the PTCDI and they represent six phases: domino, herringbone, brick wall, porous, waveform and canted. The canted, waveform and domino phases were found to be the most energetically favourable. The geometrical characteristics of the canted, brick wall and domino phases calculated in the gas phase match very closely those observed on the Au(111) surface experimentally by the STM method. There is another point worth mentioning: one dimer connection was found theoretically to be two times stronger than the best PTCDA dimer [15]. Nevertheless, in our calculations the best 2D structure that has been found (the canted monolayer) has only slightly higher binding energy as compared with the best PTCDA monolayer, which is unexpected. This result clearly demonstrates the importance of all the dimer connections involved in the binding of the monolayer: sometimes, in order to accommodate the strong bonds, weak connections must also be used, and this may reduce the overall stability of the assembly.

In order to have a complete characterisation of molecule–molecule interactions, one may need to take into account the effect of the vdW dispersion forces in the assembly of the monolayers. The vdW-DF method does not affect significantly the molecular systems connected by strong hydrogen bonds; however, we find that weaker hydrogen bonds may be significantly strengthened by dispersion interactions. As expected, the geometries of these systems remain the same as calculated with the PBE functional, and the order of the dimers with respect to their binding strength does not change either.

Chapter 4 of this thesis is devoted to the description of the molecule-surface interaction. Three points have been addressed: the binding (adsorption) energies of the flat molecules to the Au(111) surface, the corrugation of the energy across the surface and the commensurability between the lattice vectors of the melamine network and those of the Au(111).

The first point is concerned with the importance of the vdW interaction between molecules and the surface, which is missing in standard DFT functionals. We find that the binding (adsorption) energies of flat molecules (such as melamine, PTCDA,

PTCDI, NTCDA and NTCDI) to the Au(111) surface are affected significantly by the dispersion interaction. While the PBE functional predicts very weak binding which does not depend on the lateral sizes of the molecules, the dispersion interaction, as implemented in the vdW-DF method, drastically changes this picture: the binding energies increase many times with the dispersion interaction and they provide most of the binding. We find that the binding energy depends linearly on the size of the planar molecules, i.e. larger molecules experience a larger vdW interaction to the gold surface. This is in agreement with the available experimental data and intuitive understanding that the vdW interaction is additive by nature. In addition, considering a melamine-melamine dimer adsorbed on the gold surface, the vdW interaction does not effect the hydrogen bonding within the dimer and only the adsorption energy increases dramatically.

Planar molecules such as melamine, NTCDA, and PTCDA, as evidenced by our GGA calculations, show an extremely small corrugation of PES as they are absorbed on the surface. This suggests that single molecules must be very mobile during their deposition at room temperature. The vdW interaction provides significant binding to the surface; the flatness of the surface potential still remains in place, i.e. the molecules cannot easily escape from the surface, but can freely move around. In other words, these molecules on the Au(111) surface may be considered as a 2D gas.

The other problem considered is related to the *incommensurability* of the lattices of gas-phase networks of adsorbed molecules and of the corresponding substrate. The problem here is not simply computational; indeed, when placing molecules on the surface and applying periodic boundary conditions, one is constrained by the chosen supercell of the substrate. It is difficult to decide, on energetic grounds, which supercell periodicity is the most favourable as different choices may give a different number of adsorbed molecules per number of surface unit cells. In reality, the problem is even more complex: either the actual periodicity of the molecular network becomes very large (tens of lattice vectors of the surface in the repeating unit) or there is no periodicity at all, i.e. the network is never truly periodic and it can only be considered approximately periodic. The DFT calculations of chains and clusters of melamine molecules on the gold surface showed that there are variations of the inter-molecular distances and angles in these systems when compared to the gas phase. These variations are usually small, but they may be considerable in some cases. Thus, even though a network may look periodic in the STM images, it may in fact be slightly aperiodic, i.e. distorted, because of a subtle effect of the surface. The mismatch between the gas-phase lattice of the molecules and that of the substrate may be the reason for the formation of complex assemblies consisting of domains of finite islands with intermediate "domain walls" regions [1] serving to relieve the strain in the monolayers.

In Chap. 5 we analysed the presence of bright spots seen in the STM images of the assembly of melamine or some other organic molecules on the Au(111) surface. Our assumption is that gold atoms are trapped in the network and that the deposition rate of the molecules and the resulting coverage play an important role in the formation of the network itself. In order to prove our hypothesis we used different approaches, performing calculations with both PBE and the dispersion-corrected PBE methods, and studying possible kinetic mechanism which may be responsible for trapping gold atoms.

A detailed analysis of the kinetic mechanisms allowed us to suggest that at small deposition rate gold atoms, which are always present on the surface and quite mobile on it, would quickly form blocks or complexes with the melamine molecules, and these can also diffuse freely on the surface. When a second melamine molecule comes into proximity of these blocks, a hexagon formation is initiated. Diffusion calculations show that if a melamine approaches a "melamine+Au ad-atom" block from a direction different to the position of the Au atom, a hexagon may eventually be formed at RT with the Au atom either inside or outside of the hexagon. However, even in the case of the Au atom appearing outside the hexagon, when additional molecules come to form a second hexagon attached to the first one, this Au atom would end up in the centre of the second hexagon.

These results highlight the complexity of the process of formation of a hexagonal network with Au atom trapped inside it. In order to investigate in more detail the mechanism of the formation of the hexagonal network with Au atoms, we should perform calculations of the energy barriers for several elementary processes involved using e.g. the NEB method [16–19]. The obtained results will be used to set up the parameters to perform Kinetic Monte Carlo (KMC) [20] calculations which could characterise the exact dynamics of the process.

In Chap. 6 we considered the assembly of N-aryl-modified nucleobases adenine, cytosine, guanine and thymine (DNA derivatives) on the Au(111) surface. All gas phase structures have been considered with our systematic approach, as it has been used for the other organic molecules considered in this thesis. We have also characterised single molecules and some dimers on the surface with our vdW-DF method and modelled the STM images for single molecules and the pairs using the Nt_STM code. These tools have proved to be useful in identifying the shape of the molecules in the experimental STM image and the periodicity of the network formed by the assembly of these molecules. The comparison of our models with the experimental STM images demonstrated an excellent agreement.

In our characterisation of the self-assembly of flat molecules on the Au(111) we successfully reached a few goals such as the analysis of the interaction and mobility of these organic flat molecule and their small clusters on the Au(111) surface. By taking into account the vdW forces, with the vdW-DF method, we obtained reasonable adsorption energies which are in agreement with the experiments. However, the Au(111) surface is somewhat special as there is no charge transfer between the molecules and the surface. It would be of utmost interest now to extend this study to other surfaces, such as Ag(111), Cu(111) or the Ag–Si(111) surface, which may initiate stronger binding via a charge transfer mechanism. The role of vdW interaction in these cases in not yet fully understood.

Moreover, we plan to complete the analysis of the bright spots on the Au(111) surface to verify that our assumptions are correct: we want to perform calculations with the NEB method and then, using an appropriate KMC method, to characterise in detail the dynamics of the formation of the melamine network with trapped gold atoms.

Finally, in order to describe the process of formation of the DNA derivative network, it would be of interest to study the diffusion of the molecules and their pairs on the Au(111) surface and model STM images of various structure models.

References

1. Silly F, Shaw AQ, Castell MR, Briggs GAD, Mura M, Martsinovich N, Kantorovich LN (2008) J Phys Chem C 112:11476
2. Xu W, Dong M, Gersen H, Rauls E, Vázquez-Campos S, Crego-Calama M, Reinhoudt DN, Stensgaard I, Laegsgaard E, Linderoth TR, Besenbacher F (2007) Small 3:854
3. Staniec PA, Perdigão LMA, Rogers BL, Champness NR, Beton PH (2007) J Phys Chem C 111:886
4. Gabriel M, Stöhr M, Möller M (2002) Appl phys A 74:303
5. Wanger Th, Bannani A, Bobisch C, Karacuban H, Moller R (2007) Condens Matter 19:056009
6. Kilian L, Hauschild A, Temiroc R, Soubatch S, Scholl A, Reinert F, Lee T-L, Tautz FS, Sokolowski M, Ulbach E (2008) Phys Rev Lett 100:136103
7. Glockler K, Seidel C, Sokolowski M, Umbach E, Bohringer M, Berndt R, Schneider W-D (1998) Surf Sci 405:1–20
8. Chinzhov I, Kahn A, Scoles G (2000) J Cryst Growth 208:449
9. Mannsfeld S, Toerker M, Schmitz-Hubsch T, Sellam F, Fritz T, Leo K (2001) Org Electron 2:121
10. Swarbrick JC, Ma J, Theobald JA, Oxtoby NS, O'Shea JN, Champness NR, Beton PH (2005) J Phys Chem B 109:12167
11. Nicoara N, Romani E, Gomez-Rodriguez JM, Martin-Gago J, Mendez J (2006) Org Electron 7:287
12. Kunstmann T, Schlarb A, Fendrich M, Wangner Th, Moller R, Hoffmann R (2005) Phys Rev B 71:121403
13. Fendrich M, Kunstmann T, Paulkowski D, Möller R (2007) Nanotechnology 18:084004
14. Lauffer P, Emtsev KV, Graupner R, Seyller T, Ley L (2008) phys status solidi 245:2064
15. Mura M, Sun X, Jonkman HT, Silly F, Briggs GAD, Castell MR, Kantorovich L Experimental and theoretical analysis of h-bonded supramolecular assemblies of ptcda molecules on the au(111) surface. Phys Rev B (in print)
16. Mills G, Jonsson H (1994) Phys Rev Lett 72:1124
17. Mills G, Jonsson H, Schenter G (1995) Surf Sci 324:305
18. Henkelman G, Uberuaga BP, Jonsson H (2000) J Chem Phys 113(22):9901–9904
19. Henkelman G, Jonsson H (2000) J Chem Phys 113(22):9978–9985
20. Gillespie DT (2007) Stochasti simulation of chemical kinetics. Annu Rev Phys Chem 58:35–55

Index

M. Mura, *Self-Assembly of Flat Organic Molecules on Metal Surfaces*,
Springer Theses, DOI: 10.1007/978-3-642-30325-8,

MIX
Papier aus verantwortungsvollen Quellen
Paper from responsible sources
FSC® C105338

If you have any concerns about our products,
you can contact us on
ProductSafety@springernature.com

In case Publisher is established outside the EU,
the EU authorized representative is:
Springer Nature Customer Service Center GmbH
Europaplatz 3, 69115 Heidelberg, Germany

Printed by Libri Plureos GmbH
in Hamburg, Germany